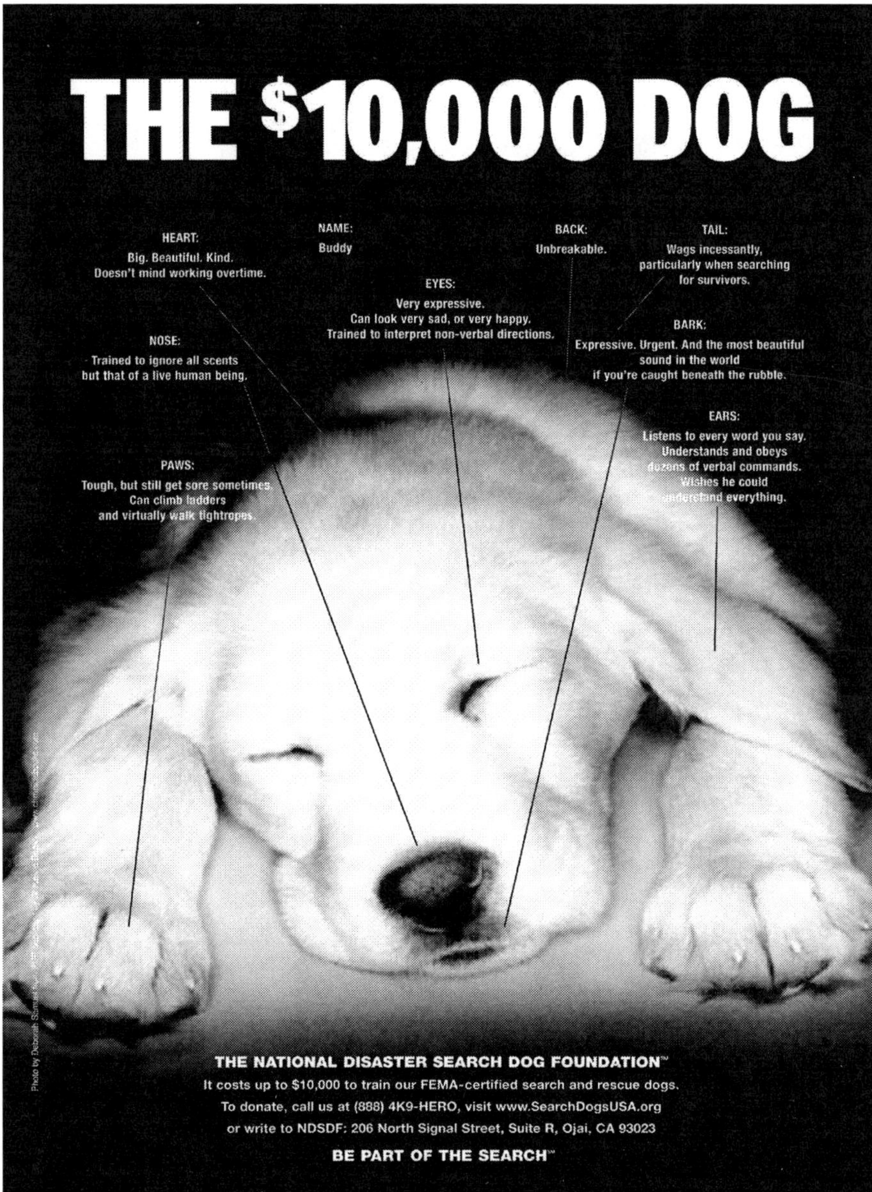
THE $10,000 DOG
HEART:
Big. Beautiful. Kind.
Doesn't mind working overtime.
NAME:
Buddy
BACK:
Unbreakable.
TAIL:
Wags incessantly,
particularly when searching
for survivors.
EYES:
Very expressive.
Can look very sad, or very happy.
Trained to interpret non-verbal directions.
NOSE:
Trained to ignore all scents
but that of a live human being.
BARK:
Expressive. Urgent. And the most beautiful
sound in the world
if you're caught beneath the rubble.
EARS:
Listens to every word you say.
Understands and obeys
dozens of verbal commands.
Wishes he could
understand everything.
PAWS:
Tough, but still get sore sometimes.
Can climb ladders
and virtually walk tightropes.
THE NATIONAL DISASTER SEARCH DOG FOUNDATION™
It costs up to $10,000 to train our FEMA-certified search and rescue dogs.
To donate, call us at (888) 4K9-HERO, visit www.SearchDogsUSA.org
or write to NDSDF: 206 North Signal Street, Suite R, Ojai, CA 93023
BE PART OF THE SEARCH™

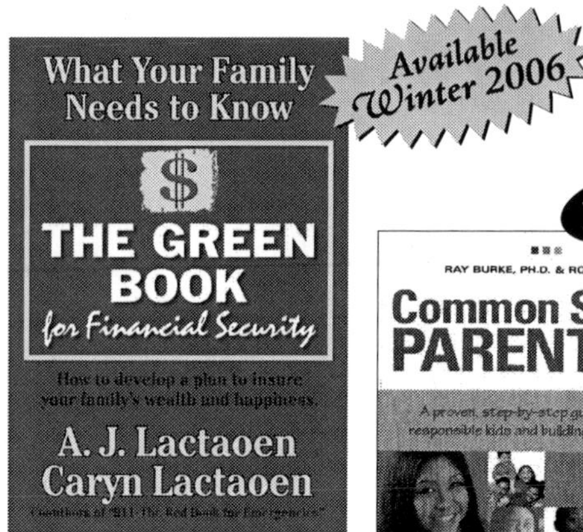

What Your Family Needs to Know

Each book has a unique perspective on matters that deal with crucial aspects of our daily lives. These are topics which are usually not taught in school or at home. Each book focuses on how to keep important things important by presenting practical, simple, well-illustrated, proven principles which anyone can apply.

Why isn't common sense common practice? Find the solution to everyday challenges in *Common Sense Principles*.

Most people want to put their family first, but life's many distractions push and pull us into so many different directions that the most important things are often neglected. You will find the inspiration to help you balance family, finances, and career in the invaluable information presented in *The Green Book for Financial Security*.

Are you a parent? You'll find comfort in *Common Sense Parenting*.

Look for the continuing series of books published and/or distributed by Legacy Books, an imprint of The Independents Group Press, in your favorite bookstore or at *www.whatyourfamilyneedstoknow.com*.

What Your Family Needs to Know

911
THE RED BOOK
for Emergencies

How to develop a plan to insure your family's safety and happiness.

"A Common Sense Approach"

A. J. Lactaoen
Caryn Lactaoen

911- The Red Book for Emergencies
What Your Family Needs to Know

Legacy Books / Independents Group Press
California 1 Oroville Irvine, California 92602
Hawaii P.O. Box 3650 Lihue, HI 96766

Ordering Information
LEGACY BOOKS are available at special quantity discounts to use as fundraisers, premiums, gifts or for use in corporate sales promotions. For more information, please call: Toll free 888-245-4604 or email: specsale@whatyourfamilyneedstoknow.com
ISBN 0-9771577-7-6 SAN 256-8578

Publisher's Cataloging-in-Publication Data
Prepared by Quality Books Inc.

Lactaoen, A.J.
911, the red book for emergencies: what your family needs to know: how to develop a plan to insure your family's safety and happiness/ A. J. Lactaoen, Caryn Lactaoen.
p. cm
Includes bibliographical references and index.
"A common sense approach."
ISBN 0-9771577-7-6
1. Home accidents-Prevention. 2. Safety Education.
I. Lactaoen,Caryn. II. Title.
TX150.L33 2005 613.6
QB105-200172

Credits
Editor: Joyce Obenauer
Design & Layout: Harumi Kobayashi of Mejiro Graphics
"Watch Your Back" Cartoons: Mario Macari

Printed in the United States of America
10 9 8 7 6 5 4 3 2 1

Dedication

For our grandchildren
Andrew Jon and Lauren Elizabeth
— with the hope that their new generation
will be able to shape a safer and kinder world.

"Parents are the best thing
between their children and harm's way."

——— *AJL*

Warning – Disclaimer

This book is designed to provide information about recommendations on how to handle emergency situations which threaten the safety of family members. It is sold with the understanding that the author and publisher are not engaged in offering legal, medical, accounting or other professional services. If legal or other expert assistance is required, the services of a licensed, certified professional should be acquired.

It is not the purpose of this guidebook to provide all the information necessary for emergencies, but instead to simplify, and complement other available data. Readers are encouraged to refer to the many resources listed in the Appendix.

Every effort has been made to make this manual as accurate and complete as possible. However, there is always the possibility of inadvertent mistakes, both typographical and in content. Furthermore, this manual contains information that is current only up to the printing date. Please refer to future editions for updated material.

The primary purpose of this book is to educate. The authors and Legacy Books (TIG) shall have neither liability nor responsibility to any person or entity with respect to any loss or damage caused, directly or indirectly, by the information contained in this book.

Contents

About the Authors 9
Author's Note 11
Acknowledgments 13
How to Use This Book 17
Emergency Response Survey 21

Introduction 29
Because We Care 31
Common Sense Approach 34
Commons Sense Principles for Emergencies 44

PART 1 Communications 47
The Basic Elements 48
The Challenges 49
The Four Categories 64
The Keys to Understanding 66
Passwords 68

PART 2 Multiple Dangers / Multiple Risks 71
Medical Emergencies 77
Accidents 84
Missing Persons 92
Victims 101
Fire 122
Natural Disasters 126
Terrorism 155

PART 3 Quick Reference 161
Safety Tips at a Glance 163

PART 4 Personal Stories 195
From the Heart.... 197
Share Your Own Stories...... 234

PART 5 Interviews 237
Deidre Downs, Miss America, a D.A.R.E. Graduate 239
Tom Hamilton, father of Bethany Hamilton, shark attack victim...242

Doug Hart, "How We Got Our Daughter Home Safely" 244
Sandy Henry, survivor of deadly Oklahoma City Tornado 249
Michael Hingson & Roselle, PR for Guide Dogs for the Blind 254
Kristine Lesperance, PhD., K-9 Search and Rescue Instructor 260
Doug Ludolph, Fire Battalion Chief, Los Angels County 265
Chris Schwart, Morgan Stanley, World Trade Center 268
Darrell Scott, father of Rachel Scott, Columbine School victim 272

PART 6 **Questions, Quizzes, Likely Scenarios** 279
Difficult Questions - Valuable Answers 281
Sample Quizzes to Prepare for the Real Challenge 283
What If... Could Be Just a Heartbeat Away 290

PART 7 **Resources** 303
Organizations that Inspire 304
Companies that Care 305
Volunteer Groups that Contribute 307
Government Agencies 310
Useful Websites 311

Afterword 313
Hindsight Teaches Us Foresight 314

Appendix A
Schools 319

Appendix B
Sample Emergency Kits 327

List of Illustrations 331

Notes / References 333

Order Forms 335

Back of the Book: Your Personal Family Profile 337
Directory 339
Forms 340

About The Authors

A. J. Lactaoen

A. J. Lactaoen is an author and motivational speaker whose mission is to promote the principle of teaching people to help themselves. He has been a college professor, a producer, an entertainer, a sports editor, a newspaper columnist, a business owner and a special events organizer. He graduated with honors from the University of Hawaii and did his masters and doctoral studies at the University of Minnesota. He is the founder of 911 Seminars and is a certified instructor in emergency training for First Aid, CPR and AED. A. J. is a member of the National Association of Search & Rescue and a volunteer Community Emergency Response Team member. He is the author of the *Precepts of Happiness* and *Common Sense Principles*.

Caryn Lactaoen

Mother, teacher and successful business entrepreneur, Caryn along with A. J. recently became a proud grandparent of twins. Caryn taught in the public and private sectors for many years then decided to enter the world of commerce as a flower broker where she quickly established a clientele list which included the Neiman Marcus chain, the Saks Fifth Avenue stores, Adrian Vittadini, Sony, Polo and many first class tour companies.

Since 9/11, Caryn has joined her husband in focusing on their book series "What Your Family Needs to Know" and their workshops in "Saving Lives Through Education." She is also certified as a trainer to teach "Common Sense Parenting," a program developed by the renown Girls and Boys Town organization in the heartland of Nebraska.

Saving Lives
Through Education

911 SEMINARS

We Provide Workshops & Training Programs for:

Families: We teach Common Sense Principles to help families prepare for any emergency situation and to expect the unexpected.

Schools: We work with faculty and staff so they are prepared to deal with students and parents in every type of emergency.

Businesses: We provide training to employees so they are better prepared to deal with customers and clients during a disaster. We also furnish management with the tools to maximize employee performance and reduce absenteeism.

Authors' Note

Our special thanks go to all the wonderful people we met while traveling around the country gathering information for "What Your Family Needs to Know – The RED Book for Emergencies." We discovered early on that we could learn more by just engaging people in casual conversation rather than through the formal interview process. We did some of the usual interviews too as you will find in Part V of our Q&A sessions.

We talked to cab drivers, waiters, concierge, bell hops, maids and security personnel. We visited with the firemen in their stations in Chicago, police on their walking beats in Brooklyn, customers in bookstores and coffee shops, business owners in Atlanta, vacationers on Mille Lacs Lake in Minnesota, guests and bartenders in the nation's Capitol and even shared Christmas cookies with bank tellers and Santa in Park Ridge, Illinois. We visited with people in Miami a few weeks after the onslaught of the worst hurricane season in recent memory. We shared a common bond with the visitors who came to see the revered site of the Twin Towers in New York and the reconstruction of the Pentagon.

From the heartland of Boys Town, Nebraska to Orange County, California, up to Las Vegas and west to our home in Hawaii, almost everybody we talked to expressed the feeling that for them the world had changed on 9/11. Then they stressed that they were more aware now, more appreciative of their family and friends. And they noticed that more people seemed to be looking out for one another. Here's an example from a conversation we had with a room attendant at the Waldorf –Astoria in New York city. "You know, I was sick one day and couldn't report to work. Well, you wouldn't believe it," she said grinning mischievously from ear to ear, "but my boss called and talked to my husband." She paused to make sure we understood how unusual this was. "You know why he called?," she asked. Not waiting for an answer she proudly continued, "he just called to see if I was o.k.? Imagine that!"

That positive sentiment reinforces a quote we saw carved in stone on an historic building in Washington, D.C.

In essentials, unity.
In non-essentials, liberty.
In all things, charity.

——————— *Richard Baxter*

Acknowledgments

"Feeling gratitude and not expressing it is like wrapping a gift and not giving it." —————— *Unknown*

We would like to express our deep appreciation to:

The American Red Cross Hawaii Chapter; Alfred Darling, ARC Kauai Branch Manager; Patsy Hamada, ARC Trainer; and our fellow instructors in CPR and First Aid - Walter Berry, Naomi Nordmeier, Maile Poai, Karen and Scott Poland, Tom Schreiber, Nick Solatre, and Scott Theis.

The Community Emergency Response Team (CERT) members with whom we shared many hours of training by Instructor Gerald Shinn; team leader Sal Ochoa, Tony Allen, John Burns, Rick Drayton, Donna Henrickson, Murray Hudson, Elizabeth Kendall, Kay Key, Bette & Jack Layton, Barbara & Ralph Leaman, Tima Loidl, Janice & Keith Olsen, Debi Rickert, Sandy Swift, Janet Teves. Gary Pacheco, and Susie Vance.

Search and Rescue volunteers Debra Gochros, Peter Kahapea, Ginny Baldwin, Alvin Castelo, Chad Pacheco, Lalei Suzuki, Juliet Montcrief, and Kuma Davis for helping us survive the night mission in Waipake and passing the grueling tests given by Kris and Ken Lesperance our SAR Instructors.

Firefighter Mason Chock and Drew Stenovich - our energetic "Ropes Course" Instructors.

Lieutenant Miles Tanabe, Pamela Cournoyer, Kathy Hyde, Roy

Kindrick and Jimmy Sanborn, the training team from the Western Community Policing Center for sharing their expertise in COPS (Community Oriented Policing Services), U.S. Department of Justice.

Mark Marshall, Civil Defense Island of Kauai, for his lectures on Tsunamis; and the following for providing important information - James Weyman, Director, Central Pacific Hurricane Center; Jason Moats, Emergency Services Training Institute, The Texas A&M University System; Robert Hansmeier, Transportation Security Administration, U.S. Department of Home Security.

The many dedicated professionals who are devoting their time and expertise on "Bioterrorism" to teach First Responders and emergency volunteers – these include Scott Lillibridge, MD - University of Texas; Mary Desvignes-Kendrick, MD, MPH - University of Texas; Robert Emery, DrPH - University of Texas; John Herbold, DVM, MPH, PhD - University of Texas; David Lakey, MD - University of Texas; Stephen Waring, DVM, PhD - University of Texas; Eric K. Noji, MD, MPH - Office of the U.S. Secretary of Defense; Dan Boatright PhD, MPH - University of Oklahoma Health Sciences University; Brenda Elledge, DrPh, RT - University of Oklahoma Health Sciences University; Trudi Nekomoto, MS - Texas Department of Health; Kate Wright, PhD -Saint Louis University; Mike Thomas, PhD - Saint Louis University; Mark Keim, MD - Center for Disease Control and Prevention; Ray Swienton, MD - University of Texas Southwestern Medical Center.

Edward Fernandez of Safefood Solutions who taught us the importance of temperature and the dangers of improper handling of foods.

Harry Hoots, professional security manager and our instructor in CDT (Compliance, Direction and Takedown)

To emergency First Responders everywhere— firefighters, police, medical personnel, search & rescue crews, volunteers — and to all the members of our Armed Forces at home and abroad — thank you for making our world a safer place.

For their inestimable contributions to the project, we also extend our thanks to the following:

Tom and Debbie Niblick from The Printmaker for their brainstorming ideas for our logo.

Mario Marconi for his genius of creating pictures to describe the ideas I gave to him in words.

Riley Baker, our virtual assistant, who is always ready to work no matter what the project or assignment.

Sheryl Sarkoff of Computerweb for her help in creating the impressive website for "What Your Family Needs to Know."

Andy Stennett for his endeavors in putting together our slide programs to music on DVD format.

All the people who agreed to be interviewed and to share their comments with us including Deidre Downs, Tom Hamilton, Doug Hart, Sandy Henry, Michael Hingson, Kristine Lesperance, Doug Ludolph, Chris Schwart, and Darrell Scott.

Harumi Kobayashi, our extraordinary artist of Mejiro Graphics, who spent countless hours laying out the book in a clear and attractive way. Kudos for your wonderful talent, Harumi.

Special thanks go to our editor Joyce Obenauer and to our publisher Richard Obenauer of Legacy Books, an imprint of the Independents Group Press for their belief and untiring support of our projects and for their friendship.

Celeste and Paul O' Brien for their insight from a school administrator's point of view.

Cristie and John Rubeo and daughters Maggie, Lanie, Stephanie; Sally and Brian Miller and sons Ben and Ayden; Fran and Elliott Higgins and their daughter Eva; Betty and Bill Charney and their twins Aaron and Anna – for providing us a view of the world from a young family's perspective.

Carol Ann for her contagious enthusiasm and encouragement.

Tom Norberg and Howard Lundy for showing us the devastation of Florida's hurricanes and introducing us to the indomitable spirit of the people who live and work there in spite of Mother Nature's fury...

Koa for showing us how to survive a hurricane. Buster Brown for demonstrating how the ideal home security alarm works.

Most importantly, we would like to extend our gratitude to our wonderful sons and their spouses – Tugs and Mako; Jimmy, Nina and our beloved grandchildren, twins Drew and Lauren for the rich and meaningful life experiences we have written about.

How To Use This Book

"Each generation will reap what the former generation has sown."

——— *Anonymous Proverb*

If you're like a lot of people we know, one of your favorite places to post important phone numbers is on the door of the refrigerator using decorative magnets to hold the memo in place. Chances are, there are also pictures of the pets, favorite drawings the kids gave you and birthday cards you want to save, all arranged haphazardly. Most likely, in an emergency, either Grandma or the neighborhood babysitter won't be able to find the critical information - if it's still legible.

Remember all the brochures and flyers you picked up at the PTA meeting about very important matters like what to do in a hurricane; evacuation procedures in case of a flood; your child's school lock down procedures and so on. Do you know where they are?

Or how about all the information you downloaded from the web? You printed up articles on what to do in a natural disaster, a terrorism attack, and a power blackout. Unfortunately, you never found the time to read them.

The objective of **911 The Red Book for Emergencies** is to give you all the practical advice you need in one complete user friendly, easily recognizable format. The book is divided into individual chapters that can be read independently and in any order. You'll learn the practical strategies and behaviors that will enable you to take back control of life in a common sense way. Complex information is turned into easy-to-understand ideas and written in comfortable, everyday language.

It is part guidebook, part journal, part workbook, part standard book, part resource book. We encourage you to write in it, underline, make notes. Just remember that someone else may have to read the important information at a crucial time. **So write clearly**.

Are you ready? Let's Roll!

Emergency Response Survey:

Take the Emergency Response Survey as a benchmark of where you stand as of this moment in time. Let every member of the family take the survey. After you've read the book, take the survey again. You'll be surprised at the results.

Introduction:

Discover a life management method which uses simple tools like the POO Factor. Learn a new set of ABC's and 3 R's. Practice the Common Sense Principles for Emergencies and the Power of 3.

Communication:

Fact of life: You won't get help unless someone knows you need help. Clear communication is a basic requirement to protect your family especially in times of emergency.

Multiple Dangers/Multiple Risks:

A basic survey of the major challenges we face in today's world.

Apply the lessons learned in the introduction to increase your chances of preventing, avoiding or recovering effectively from any danger.

Quick Reference:

One page tips at a glance. Use this section to refresh your memory. Play a game with your children and let them review one page a day and then take a quiz. Let us know if you think there are topics we missed.

Personal Stories:

Here's your chance to get the grandparents to tell their stories to the little ones. You'll be amazed at how many you probably never heard before. If you think you have a story worth sharing, send it to us. Perhaps we will include it in a future edition.

Interviews:

We can learn much from other people's personal experiences. Talk to the people you respect and ask the questions that are important to you.

Questions, Quizzes, Likely Scenarios:

Imagine how much you can discuss with your children during the long drive home while stuck in bumper to bumper traffic. Plan out a likely scenario or let one of the kids draw one of the "Watch Your Back" cartoons.

Resources:

Take advantage of the information and programs described under this category. You may be surprised to find out how much help is available.

Afterword:

If we don't learn from our mistakes, then we've wasted past efforts. Don't be a shoulda, woulda person. Not when the stakes are so high!

Appendices:

There's a wealth of information in this section especially for teachers. Find out how you can get your students involved in scholarship opportunities.

Back of the Book Forms:

Fill out all the forms in this section of the book. The information contained in your family profile will be invaluable in an emergency. **911 - The Red Book for Emergencies** is easily identifiable. Keep it handy by the phone or on your desk. Tell everyone in your family where it is located. Consider keeping a duplicate copy in the car and be sure to take a copy with you when you travel.

NOTE: This is an interactive book. Talk back to us! You can e-mail, fax, write or call us with your questions, comments, suggestions and submissions.

Emergency Response Survey for Families

First,

each person should take the survey independently. Based on your own personal knowledge, without asking your spouse or anyone else in your household, answer the following questions.
If you do not know the answer, check the question mark (?). If the question is not applicable, check the not applicable (NA) box.

SEE SCORING GUIDE AT END OF SURVEY.

Next,

compare the results of your survey with the other people in your household . By combining all the scores together and figuring out the average, you will have your important ***family emergency profile***.

Part 1

COMMUNICATION

Ⓐ

In an emergency, can you contact the people or organizations listed below immediately:

	YES	NO	?	NA
Fire Department				
Police Department				
Poison Control Center				
Hospital				
Ambulance				
Doctor				
Pharmacist				
Veterinarian				
Your Child's School				
Spouse or Parents' Workplace				

Ⓑ

In an emergency, can you be reached by any of the following ASAP:

	YES	NO	?	NA
Cellphone				
Pager				
Email				
Call Forwarding				
Answering Service				
Relative				
Friend				
Neighbor				
Colleague				
Employer				

Part 2

PLANNING

	YES	NO	?	NA
Do you have an emergency plan for evacuation to the nearest shelter?				
Do you have an emergency plan for evacuation to a designated place instead of the shelter?				
Do you have an emergency plan for staying in and securing your house?				
Does everyone in the family know the plan?				
Do you know your official Civil Defense radio broadcast station?				
Do you know your child's school emergency plan?				
Do you know your workplace emergency plan?				
If you are separated, do you have a plan to make contact?				
Do you have an emergency plan while on vacation?				
Does your family have an emergency phone tree?				

Part 3

KNOWLEDGE & SKILLS

Does anyone in the family have the following knowledge or skill?

	YES	NO	?	NA
First Aid				
CPR				
Pet CPR				
Use of AED				
Water Safety				
Survival Skills				
Search & Rescue				
Self Defense Tactics				
Shortwave Radio				
Construction (electrical, carpentry, plumbing, etc.)				

Part 4

RESOURCES

Do you have any of the following?

	YES	NO	?	NA
House Insurance				
Cell Phone				
Self-powered Radio				
First Aid Kit				
Fire Extinguisher				
Smoke Detector				
Disaster Supplies				
Safety Ladder for Second Story				
Security System				
Portable Generator or Auto 12 Volt Adapter				

SCORING GUIDE

INDIVIDUAL SCORE

Number of **YES** answers × **2** points	=	
Number of **NO** answers × **0** points	=	
Number of **?** answers × **1** point	=	
Number of **NA** answers × **2** points	=	
Total Points		

FAMILY SCORE

Add every person's total points	
Number of persons taking survey	
Divide total points by the number of people in survey	

FAMILY EMERGENCY PROFILE

EXCELLENT	VERY GOOD	GOOD	OK
96–100	90–95	85–89	80–84

Twelve COMMON SENSE PRINCIPLES for Emergencies

1. The primary objective is to **prevent** any dangerous event from occurring. The second option is to **avoid** any risk of harm.
2. You may be the best chance to help someone in trouble, but if you get hurt, you can't help anyone.
3. Act decisively and without hesitation when the timing demands it.
4. Stay calm and keep your focus when the situation calls for it.
5. Ask for help from all sources when the challenge requires it.
6. You have more control over circumstances than you realize.
7. Every member of the family can contribute his/her own unique talent.
8. Minimize damages by learning the skills and sharing the knowledge needed in times of emergencies.
9. Some people may accuse you of being over protective. Protect them anyway.
10. Prepare for the unexpected. The better prepared you are, the better protected your family will be.
11. Optimize the recovery of losses by making use of all available resources.
12. Be a Good Samaritan. Do no harm and do the most good for the most people.

Introduction

"Give a man a fish and feed him for a day. Teach a man to fish and you feed him for a lifetime" ——— *Chinese Proverb*

In 1981, Adam Walsh was abducted from a Florida store and later found slain. As a way to honor his young son's memory and create a legacy for his lost child, John Walsh established the *National Center for Missing & Exploited Children*. In 1988, Walsh introduced a new program on Fox TV called *"America's Most Wanted,"* whose featured predator was caught three days later.

At almost the same time, Candy Lightner's 13-year-old daughter, Cari, was killed by a drunk driver with a long list of prior arrests. Candy turned her grief into activism and founded the organization *Mothers Against Drunk Driving (MADD)* which recently celebrated its' 25th anniversary of accomplishments.

Samantha Runnion, an innocent 5-year-old child, was abducted, raped and murdered. After Samantha's killer was captured, bereaved mother, Erin Runnion, created the *Joyful Child Foundation* as her way to give her daughter's death some purpose. The foundation organized the "Crusaders," neighborhood patrols comprised of red-vested volunteers who watch children playing outdoors or assist in walking them to and from school.

The *Amber Alert* system, which is now operating in all 50 states, was created in Texas after 9-year-old Amber Hagerman was kidnapped while riding her bike in Arlington, and murdered.

On July 29, 1994, 7-year-old Megan Nicole Kanka was enticed into a neighbor's house believing she could play with his puppy. The neighbor was a convicted sex offender who had already served time in prison for aggravated assault on another child. Megan was murdered but the result of her death was the passing of a nationwide law on the right to know if a convicted sex offender moves into your neighborhood. The law is simply referred to as Megan's Law.

Dr. Henya Shanun-Klein, a Houston psychologist, whose own 11-year-old daughter, Gili, was killed by a reckless driver in 1990, has emphasized,

"We don't stop being parents, we want to continually remind the world that our child existed and that she made a difference in the world." *[1]

BECAUSE WE CARE

We take inspiration from these grieving parents and the many others not included here, simply because there are too many. We should also thank these champions of issues which are important to all of us – for their persistence and for their courage in helping new laws and new regulations get passed which help to protect all of our children.

Let us also continue to be proactive in our campaign – to develop new and different strategies to protect our families. That is the purpose of our book, *911- The Red Book for Emergencies*. We want our children to grow up to be self-sufficient, contributing members of society with the keys to finding happiness. Knowledge is not enough. Understanding is not enough. What we need is commitment. We believe you will make that commitment for yourself and for your family. That's why you're reading this book. *Because You Care*.

COMMIT: Perceive, Predict, Prevent

ACT: Plan, Prepare, Practice

REPEAT: Plan, Prepare, Practice

EVALUATE : Perceive, Predict, Prevent

Note: Perceive, Predict, Prevent = Protect

People Do Stupid Things!

That's the byline of the Vonage commercials on television – "People do stupid things!" *The Fountain* shows a photographer climbing up on a fountain in an attempt to get a better angle for a wedding picture. He falls in and thrashes around in the fountain... *The Rocket* depicts a young man sending up a toy rocket which blasts off straight up into the sky. The young man is heard to say he can't see it while he's focused on staring up into space. A second later the rocket whizzes by as the idiot ducks just in the nick of time... *The Tree* starts off with the sound of a chainsaw and slowly the camera focus widens so that the viewer can see how big the tree is and that just to the left of the tree is a parked car. Yes, you're right. Mr. Paul Bunyon keeps on sawing and before you can say "Timber" the tree smashes the car... *The Batter* commercial has a teenager swinging a bat in front of a patio with glass doors... Yes, you are correct again. The baseball teen swings and misses but the bat goes flying breaking the glass into a thousand pieces...

Why do we think these imitations of *America's Funniest Videos* are so funny? We think it's because they're almost like cartoons. Nobody gets hurt, and because there's humor in the fact that these are probably real accidents that have happened.

On the other hand, when people do stupid things that result in tragedy, then our emotions flip flop from sorrow to anger then back to grief for the victims.

Some examples:

- Leaving the car running with a child in the backseat
- Tossing a cigarette out of the window during a drought in the hot summer

- Diving off a cliff head first
- Driving 85 mph in a 25 mph zone while intoxicated
- Playing with a loaded gun

Why Do People Do Stupid Things?

Here are some of the reasons. Maybe you can add to the list.

- *Fear: Paralysis*
- *Laziness: No need to check if it's safe or not.*
- *Carelessness: Rush, rush, rush*
- *Fatigue: Burn out*
  Just hope it's not your doctor, pilot, or dentist
- *Social Pressure: "What will the Jones' think ?"*
- *Addictions: "I can't help it."*
- *Ignorance: "I didn't know"*
  Ignorance is not bliss.
- *Denial: "Not Me!"*
  If you don't think it will happen to you then you won't take any precautions.
- *Blame: "Somebody told me to do it!"*
  Benjamin Franklin once said that he who is good at making excuses is seldom good for anything else.

Question: What's the difference between a smart man and a dumb man?

Absolutely nothing. They both think they know everything.

COMMON SENSE APPROACH: A Different Perspective

Our book, ***Commons Sense Principles***, deals with the topic of Common Sense Perception in detail. Because our focus here is on how to deal with emergencies, we will present an abbreviated version.

How many times have you heard someone say, "That's just plain common sense," or read in the news, "Just use your common sense." Based on these comments, it seems that people assume that everyone has or should have what is known as "common sense" and

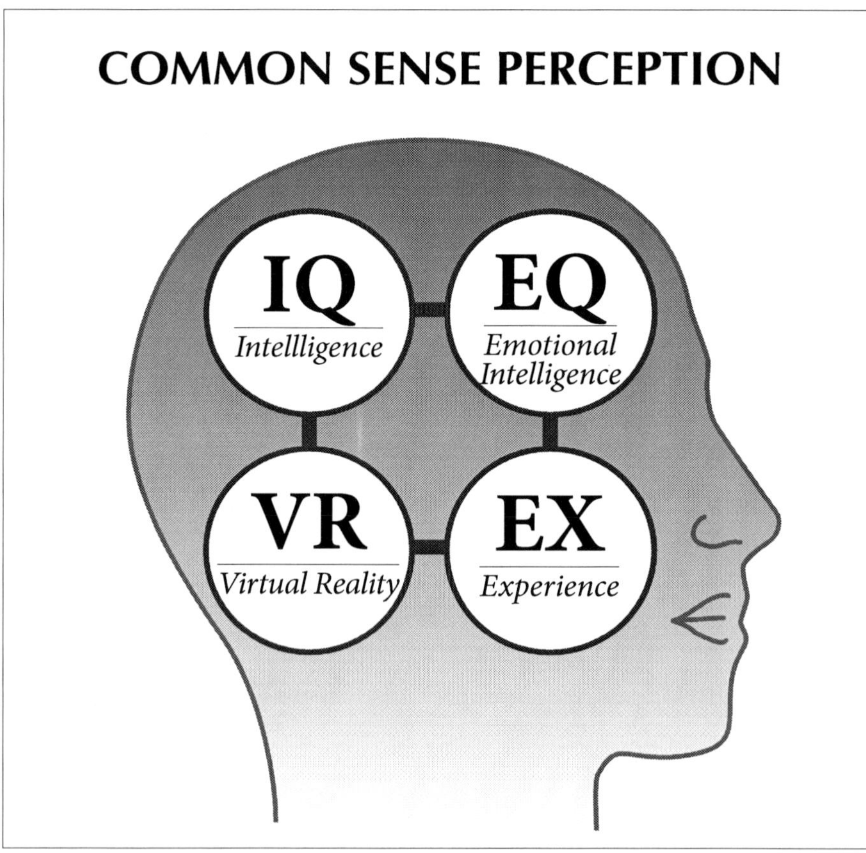

that people are born with this innate, practical wisdom. As a matter of fact, a nationwide survey reveals that the majority of Americans believe that a high level of common sense can lead to healthier, happier and more successful lives. So the question remains, why is common sense not common practice?

We believe that common sense is a life skill that has to be developed. Only a part of our intelligence is inherent in our genes. However, we have the ability to develop and enhance the rest of our *Common Sense Perception*.

See our figure head that is divided into four parts:

- IQ = Intelligence
  Einstein is an example of a genius who was extremely strong mathematically while being relatively weak verbally. Book smarts is another way to describe the analytical skills measured by IQ tests.
- EQ = Emotional Intelligence
  An individual's ability to manage him or herself as well as other relationships. It consists of four fundamental capabilities of self-awareness, self management, social awareness, and social skill. Employers now recognize that an individual with high EQ makes for good leadership potential and is a competitive edge in the marketplace.
- VR = Virtual Reality
  This is our label for learning that takes place through multiple learning styles such as visual learning, auditory learning, musical learning. For example, we learn from watching movies, by listening to recordings, by hearing stories, parables and fables read aloud to us. We learn about ethics and philosophy from proverbs

and speeches or from mentors and the example of others. It's visualizing how to dance before we actually do the dance steps.

- EX = Experience
  We learn by actually doing the deed and succeeding so we know what works; or failing so we know what doesn't work. As Marilyn vos Savant, the woman with the highest IQ ever recorded said, " If you don't make mistakes, you probably don't know much."

An individual with a highly developed common sense perception has a critical balance of all four qualities and continues to develop them regularly. Which of these are you most comfortable with? Do you consider yourself more analytical or creative? Do you think you are more practical or theoretical? The following story dramatically illustrates the difference between the two.

Two students from a prestigious Ivy League school were camping in Yellowstone Park when they came across a grizzly bear. The first student, an academic whiz with a full scholarship, immediately calculates that the bear can reach them in exactly 19 seconds.

"No way we can outrun him," he tells his companion who is putting on his running shoes. The other boy, who barely passed his entrance exams, says to his friend:

"I don't need to outrun the bear. I just need to outrun you."*[2]

As a follow up on that thought, practical intelligence versus intellectual intelligence, did you know that scientists involved in the creation of Artificial Intelligence (AI) are at a loss when it comes to instilling commonsense reasoning into it's robotic computers. The reason is the difficulty of programming emotional intelligence into their systems.

Circle of Influence

Relying on our common sense perception, we conclude that if we want our children to grow up to be self-sufficient, contributing members of society then we need to teach them that responsibility begins with the individual. Once the individual is able to take care of himself/herself, then he is able to help others in the family. As soon as the family is able to take care of their survival and safety needs, then they are able to help their neighbors. This circle keeps expanding outward in a continuing cycle of helping hands.

See illustration of Circle of Influence.

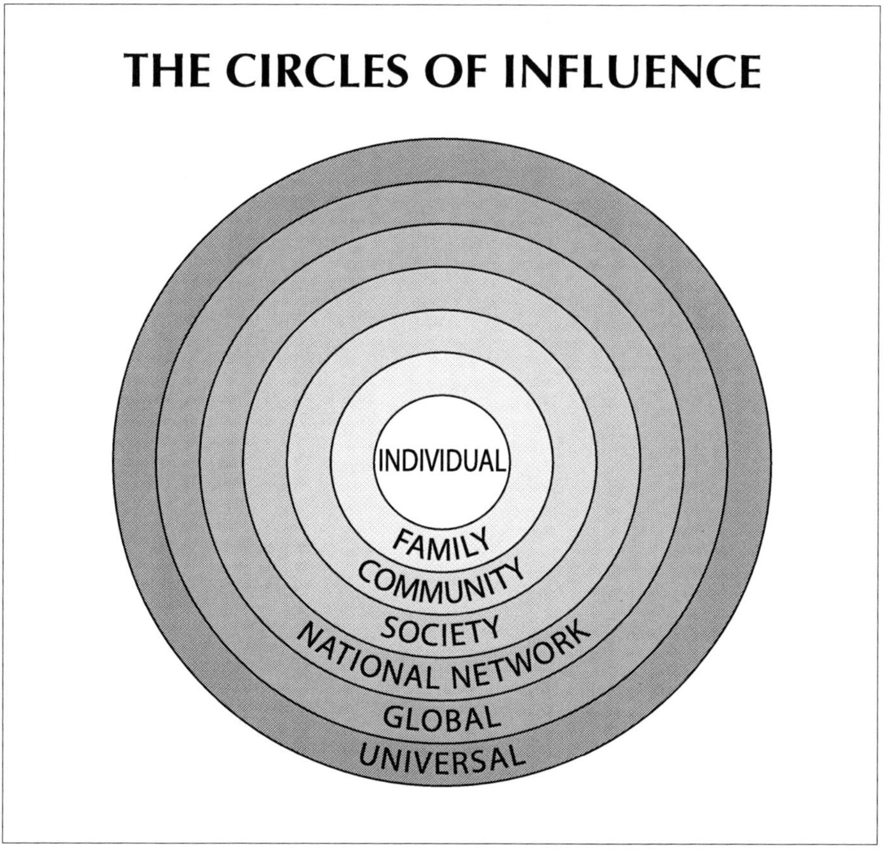

Cycle of Generations

Our goal is to provide a safe and nurturing haven for our families. We can do this so much more effectively when each generation shares in the task. Children and grandparents can create a bond that can keep a family together through the worst of disasters and the cruelest of tragedies. If each generation does its part, contributing their own talents and skills, then everyone benefits.

See illustration of Cycle of Generations.

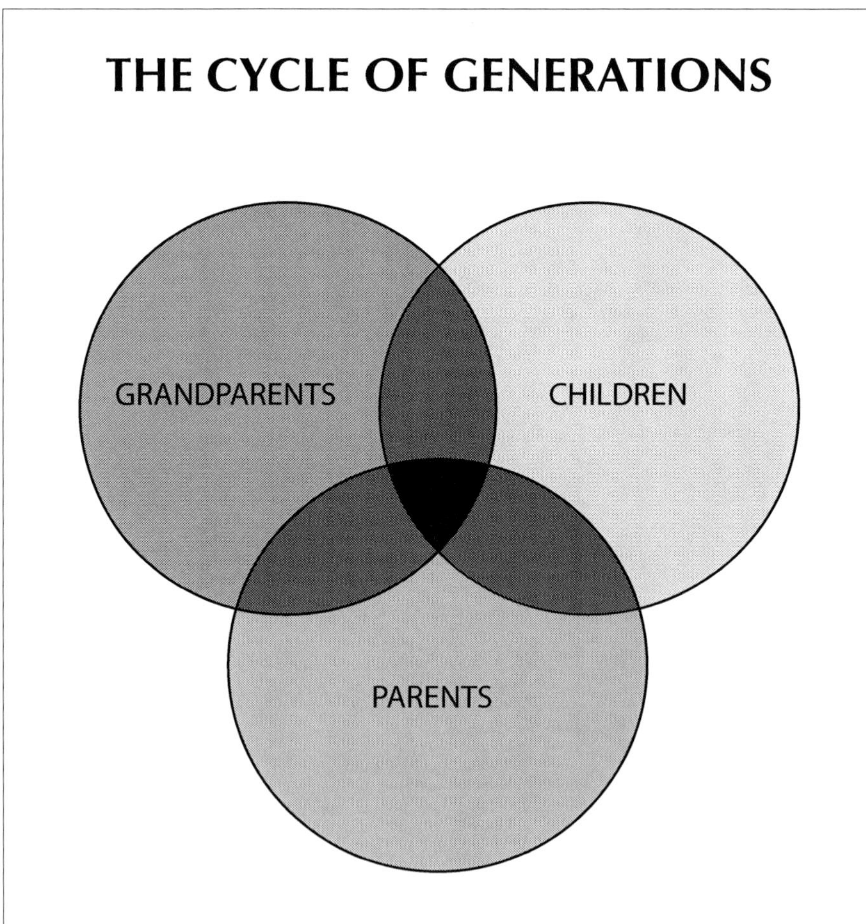

The POO Factor

The *POO Factor* is a system or method of getting back control of your life. Sometimes, we feel like we're caught on a treadmill and we can't seem to get off or stop our world from spiraling out of control.

The problem is we feel overwhelmed. We can't separate or distinguish our priorities because we don't know the art of self-evaluation. Once we graduated from school, we no longer had a teacher or mentor or a way of measuring our progress in life. The *POO Factor* provides the tool that we need to keep track of our daily attempts at improving our life. It is a self-evaluation or report card on how we're progressing. It doesn't matter what the challenge, the task or the goal might be. The *POO Factor* provides a visual map of where we are at any given moment.

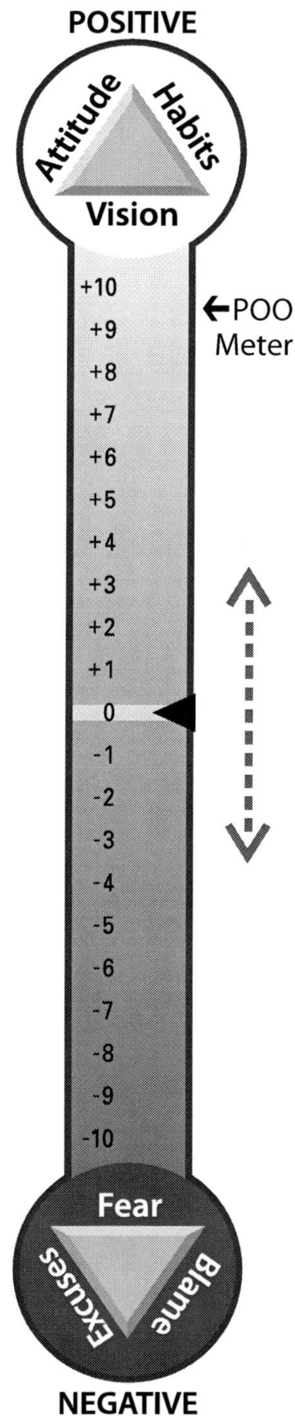

See our illustration of the *Poo Meter*:

The *Poo Meter* is a visual reminder of where you stand at any given moment on any specific goal you've set for yourself. The triangles remind us that there is a positive and a negative force at work in our lives. The arrow, which is set at zero, will move up or down. It will not stay the same. Almost nothing in life "just happens." In order for the arrow to move up, we must focus on our *Vision, Attitude and Good Habits*. Otherwise, the Law of Inertia will concede to *Fear, Blame and Excuses*.

POO is the acronym for *Probability of Occurrence*. The *POO Factor* is not a mathematical theory or a scientific experiment. It is the art of using one's highly developed common sense perception (CSP).

The chart of probabilities listed below is not the one you want to depend on.

Chart of Probabilities

BET MY LIFE	100%
I'M POSITIVE	90%
I'M SURE	80%
I'M CONVINCED	70%
A GOOD CHANCE.	60%
IT'S 50 – 50	50%
I THINK SO	40%
COULD BE	30%
MAYBE	20%
PERHAPS	10%
NO WAY	0%

So how do we improve the probabilities for our success?

Gambler or Investor ?

Do you know the difference between gambling and investing? Some people who have lost a fortune in the stock market would say there is no difference. They're wrong.

Gambling is putting your money at risk in the hopes of earning a return with no rational basis, but depending solely on luck.

Investing, on the other hand, is putting your money at risk in the rational expectation that you will earn a return.

Here's a concrete example:

What if you are dropping your 7-year-old daughter off for a sleep over at one of her classmates' house. When you arrive, the parents aren't home from work yet and you notice that one of the teens has something burning on the stove, there are candles burning in the den and you smell gas. There are old newspapers piled in a corner close to the fireplace in the living room . You remember reading that the primary causes of fires are fireplaces, unattended cooking and burning candles.

What do you do? You can gamble and let your daughter stay and hope you get lucky and she survives the night. Or you can invest your time and effort in explaining to your daughter why she cannot stay overnight. We call it Crisis Anticipation. That's what the POO Factor is all about. We can learn how to use the system as a leverage to increase the probability that you will obtain the outcome you want - to protect your family and not depend solely on luck.

If you knew that your house was going to burn, you'd prepare...
If you knew your car was going to be in a collision, you'd prepare...
If you knew you were going to be hospitalized, you'd prepare...
If you knew your house was going to be burglarized, you'd prepare...

You don't know any of these things, but yet you prepared by

buying insurance. Don't you think your children's safety is important enough to invest some time in crisis anticipation?

The Power of 3

There are three important keys in the POO Factor system and using the Leverage of Anticipation:

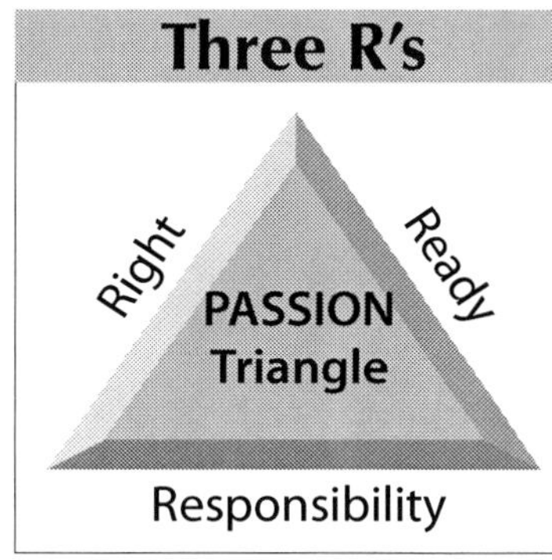

The Passion Triangle

- Responsibility:
- Right:
- Ready:

If you accept the **responsibility** for your actions, and you know you are doing the **right** thing which is consistent with your values and your beliefs, then you are **ready** to proceed.

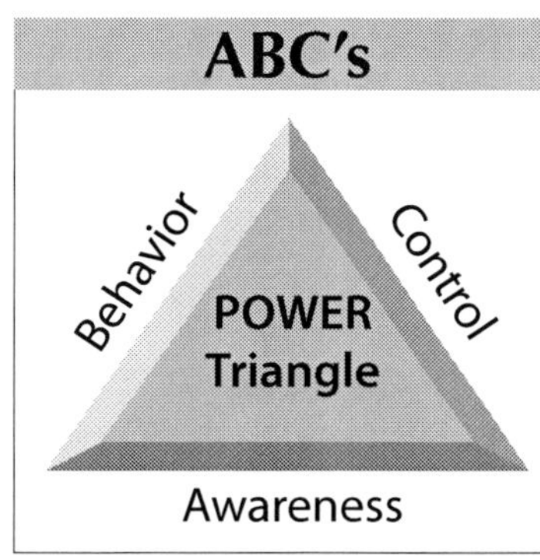

The Power Triangle

- Awareness
- Behavior
- Control

If you increase your **awareness** and change your **behavior** as needed, then you will be able to **control** your circumstances.

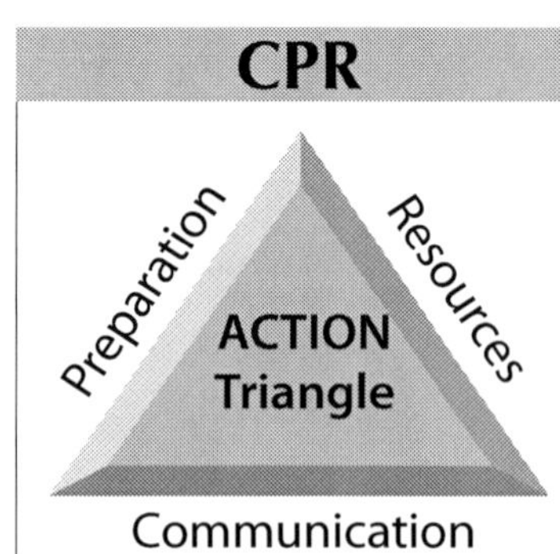

The Action Triangle

- Communication
- Preparation
- Resources

If you establish **communication** with all concerned, complete the **preparation** for your task and obtain all the **resources** you need, then you are set to go.

Mantra

No matter what the situation, always repeat the *POO Factor* mantra "What 3 things ....?" when faced with a challenge.

Immediate Danger: What 3 things do I have to be concerned about...
Imminent Danger: What 3 things do I have to be concerned about...
Continued Risks: What 3 things do I have to be concerned about...

Immediate action: What 3 things can I do to help...
Crucial Follow Up: What 3 things can I do to help...
Long Term Goals: What 3 things can I do to help...

What 3 things can you do right now to help prepare your family?

1. Follow the advice of our Homeland Security & the American Red Cross:
   Get a Kit. Make a Plan. Be Informed. Get Involved.
2. Practice the *12 Common Sense Principles for Emergencies.*
3. Be a good role model so that your kids do what you do, not what you say.

Finally, we cannot overemphasize how important Communication is in any situation. Hopefully you will share whatever you glean from this book with your family and friends. Stay safe.

fyi 1

According to the International Association of Fire Fighters, the unprecedented tragedy of September 11, 2001 claimed the lives of 344 New York City fire personnel in just one incident. Their rescue efforts saved thousands of lives. This was the largest single-event loss of life sustained by modern-day fire service personnel in war or peace times. In contrast, a total of 102 US fire fighters died nationwide in 2000.

— *Guinness World Records 2004*

What's The Family Emergency Plan?

If you're at home, at school, at work, at play or on the road, do you know what to do?

Quick Quiz

If your house is worth $ 300,000 for structure and contents and your homeowners insurance has a $ 250,000 maximum coverage, how much will your insurance company reimburse you if your home is completely destroyed in a flood?

ANSWER: Nothing. ($ 00.00) Regular homeowners insurance does not cover damages caused by floods. Special Flood Hazard Areas require special flood insurance purchased through the National Flood Insurance Program managed by FEMA.

PART 1

Communications

"Communication leads to community, that is, to understanding, intimacy and mutual valuing."

——— *Rollo May*

Proposition:

If a tree falls in the forest but there is no one there to hear the crash, is there sound?

Answer: She says – of course. He says – technically, no. Whom do you agree with?

Proposition:

If a letter is mailed but is not delivered to the person addressed, is there effective communication?

Answer: We say – no. Why? Because a basic element is missing.

THE BASIC ELEMENTS

The basic requirements for communication are:

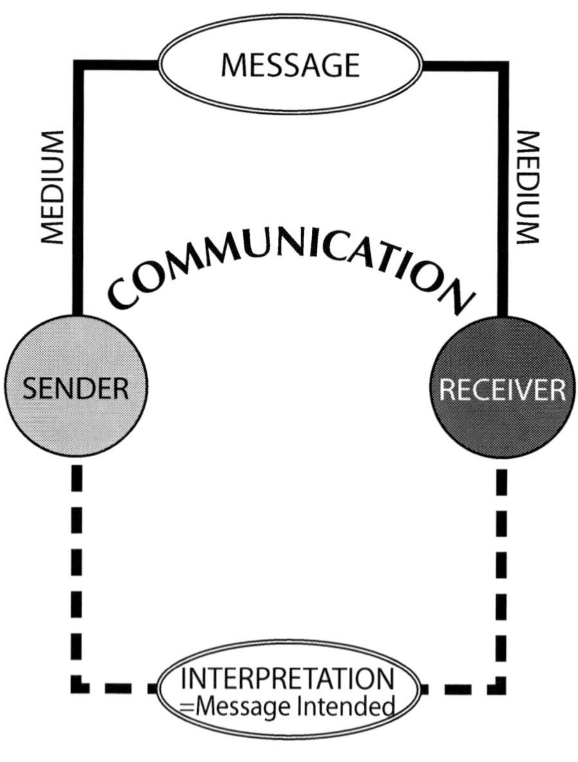

a sender

a receiver

a message

a medium

an interpretation

If the message received is interpreted in the same way the message was intended, then the communication has been executed effectively. If the message is not interpreted correctly, then there is a "miscommunication."

THE CHALLENGES

There are many reasons why effective communication is not always achieved. The sender might not be sending a clear message. Or maybe there's confusion concerning whom the message is about. Perhaps the message is relayed to the wrong receiver. Maybe the medium is faulty and the transmission is not made correctly. Also, there might be gaps in the interpretation. We've listed a few examples of each basic element of communication gone awry:

Element	Examples
Sender :	Gobbledygook; Who's on First?
Receiver :	Who's Who?; Check your Bearings
Message :	A Very Famous Zen Garden
Medium :	Tower of Babel; The 9/11 Commission Report
Interpretation :	Generation Gap; Gender Gap: Culture Gap; Organization Gap

SENDER Gobbledygook

The word "Gobbledygook" was coined back in the 1940's by a Texas congressman who compared the rambling doublespeak of politicians to the senseless gobbling of turkeys.

Semanticist Stuart Chase defines "Gobbledygook" as the practice of "using two or three or ten words in the place of one, or using a five syllable word where a single syllable would suffice." Perhaps you've had the opportunity to hear the spin masters intone "member of vertical transportation corps" in place of elevator operator; or "career associate scanning professional" for checkout clerk; how about" permanent pre-hostility" instead of simply saying peace.

The National Council of Teachers of English*[1] list the following in their Doublespeak file under the headings Job Jargon; Socialese and Dollar Bull. You can guess which is which.

The SPIN	The REALITY
urban transportation specialist	cab driver
clothing refresher	dry cleaner
economically non affluent	poor
social expression products	greeting cards
negative investment increment	loss
genuine imitation	fake
interdental stimulator	toothpick
revenue enhancement	tax increase
entropy control engineer	janitor
auto dismantler and recycler	junk dealer

The following anecdote about an incident which happened during the Lincoln-Douglas Presidential debates150 years ago, illustrates Lincoln's appeal to the simple folk.

After a long, winded speech by Stephen Douglas,

Lincoln asked the people gathered round, "How many legs would a horse have if you called it's tail a leg?"

"Five," shouted some in the audience.

"Four," Lincoln responded immediately. "Calling a tail a leg doesn't make it true."

Sadly, it's true that "Gobbledygook" has become so intrusive in our society that we now have to make laws against it. Consider the bill which was recently signed into law in the state of Florida requiring insurers to explain their coverage in plain English. Governor Jeb Bush looked at his homeowners' policy and couldn't understand it. The wording on his policy was so confusing he stated, "That isn't English." Lawmakers ironically agreed and we now have a law of the land which declares that people who buy insurance have the right to understand what their policies say. Hurray for that!

SENDER Who's on First?

Abbott & Costello's routine of Who's On First? is the classic example of confusion at it's hilarious. The unspoken realization is that the reason we find it so funny is because we sometimes do the same thing in our everyday conversation. For those of you who are too young to remember, and for the many who are old enough not to forget, here's an excerpt of the conversation:

Costello: *Are you the manager?*
Abbott: *Yes.*
Costello: *You gonna be the coach too?*
Abbott: *Yes.*
Costello: *And you don't know the players' names?*
Abbott: *Well I should.*
Costello: *Well then who's on first?*
Abbott: *Yes*
Costello: *I mean the fellow's name.*
Abbott: *Who.*
Costello: *The guy on first.*
Abbott: *Who.*
Costello: *The first baseman.*
Abbott: *Who.*
Costello: *The guy playing...*
Abbott: *Who is on first!*

Costello: *I'm asking you who's on first.*
Abbott: *That's the man's name.*
Costello: *That's who's name?*
Abbott: *Yes.*
Costello: *Well go ahead and tell me.*
Abbott:*That's it.*
Costello: *That's who?*
Abbott: *Yes. PAUSE*

RECEIVER Check your Bearings

The following story illustrates the importance of making sure you know with whom you are communicating:

First voice: *Our radar has you on a collision course with us. You should alter course 10 degrees south.*
Second voice: *We have you on our radar. Suggest you alter course 10 degrees north.*
First voice: *This is the captain of a US Navy ship. I say again, divert YOUR course.*
Second voice: *No, I say again, you divert YOUR course.*
First voice: *Sailor, this is Admiral Goodman aboard the USS Abraham Lincoln. I order you to bear 10 degrees south immediately, I am in command of a battleship!*
Second voice: *This is Seaman Farnsworth, Sir. Still suggest you bear 10 degrees north. I am in charge of the lighthouse!*

Note: It has been reported that a copy of this embarrassing conversation was recorded by the Chief of Naval Operations and the transcript was eventually leaked to the general public. Slightly different versions of this story have been floating around for 30 to 40 years. One source thinks it may have originated in a humor column in Reader's Digest. No matter how much the Navy might deny it, the story has taken on a life of it's own in faxlore and internet heaven.

RECEIVER Who's Who?

Following is another example of the importance of making sure the Sender-Receiver connection is on the right track.

A group of men are sitting in a sauna discussing business and stocks when suddenly a cellular phone rings.

"Hi honey, are you at the club?"
"Yes, dear."
"Honey you won't believe this but I'm standing in front of Giovanni's and there's a beautiful mink on sale in the window."
"How much is it, dear?"
"They're giving it away. Only $5000. Can you believe it?"
"But you already have fur coats?"
"Please dear it's absolutely exquisite!"
"Fine, fine go ahead and buy it!"
"Thank you sweetheart. Oh, not to keep you much longer, I passed by the Mercedes dealership this morning and saw their new convertible. It was to die for! I talked to the salesman and the one in the showroom is brand new, leather seats, power everything, gold colored. What do you think?"
"Honey, come on, we already have cars!"
"You promised me that I could get a convertible!"
"How much is it?"
"You won't believe it but he said he'd let us have it for $85,000 fully loaded

with all the options!!!"

"OK, OK, go ahead and purchase it!"

"I love you, you're the best husband a wife could ask for. I hope I'm not pushing it, but remember our trip we took to Paris? Remember the Brown's place with the swimming pool, tennis courts? It's on the market to be sold. I saw it this morning at the Real Estate agency. If we bought it we would have a perfect place to stay during the cold winter months!!!"

"I had actually thought about it. You say it's on the market?"

"Really, you were actually thinking about it? Can I go make an offer on it? You know it's not listed very high, and It would be perfect for our type of lifestyle!!"

"How much is it listed at?"

"Only $1,500,000 , sweetheart. It's a steal!"

"I guess we've got money put away. Go ahead and make an offer but no more than $ 1,400,000."

"This is turning out to be a great day! Can't wait to see you later tonight to celebrate!!!"

"See you tonight dear."

The man hangs up the cellular phone and asks, "So, who's phone is this?"

MESSAGE The Hidden Message

We recently visited Japan with our son Jon and his wife Makiko whose parents live in Tokyo. One of the most interesting things we did was a trip by bullet train to the historic city of Kyoto, resplendent with temples, shrines and castles. Our favorite tour was the famous Zen garden called "The Dry Garden of Ryoanji Temple."

For centuries, this simple rectangular garden filled only with sand, small pebbles and several big rocks has mesmerized millions of visitors. A diagram of the rock garden is sketched here.

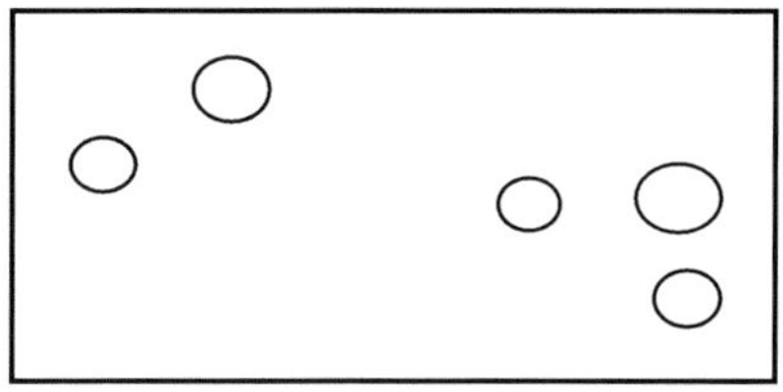

○=Rock Clumps

Fifteen rocks of different sizes are strategically arranged in five different clumps throughout the garden filled otherwise only with white pebbles. There is a platform built under roof for visitors to meditate. The intriguing attraction of the garden is that no matter where you stand, you can only see fourteen of the rocks. The final rock can only be imagined with the mind's eye.

What was the cryptic message that the creator of the garden was leaving behind for posterity? It is this mystery, we believe that draws people to the garden. We watched as people sat in silence meditating for long periods of time, trying to come to some understanding as to what the meaning of the invisible rock could be.

We took turns explaining our thoughts. A.J. offered that perhaps the message was that the reason we cannot understand life fully is because of the missing piece to the puzzle. Caryn's observation was that the only perspective from which one can see all 15 of the rocks is from the heavens. Jon suggested that everyone was able to interpret the meaning of the garden in their own unique way. Makiko summed it up by concluding that it wasn't necessary to understand everything all the time.

Life is a mystery, an enigma – that may very well be the message hidden in the sacred stones.

MEDIUM The Tower of Babel

According to the narrative in Genesis Chapter 11: 7-9, God confused the people's tongues so that each spoke a different language and they could no longer communicate with one another.

"...And therefore the name thereof was called Babel, because there the language of the whole earth was confounded; and from thence the Lord scattered them abroad upon the face of all countries."

It is estimated that there are approximately 6,500 living languages.*2

Here is a list of top 10 most widely spoken languages in the world:*3

1. Mandarin	1 billion +
2. English	508 million
3. Hindustani	497 million
4. Spanish	392 million
5. Russian	277 million
6. Arabic	246 million
7. Bengali	211 million
8. Portuguese	191 million
9. Malay-Indonesian	159 million
10. French	129 million

Surprised? You don't see some of the languages you expected, do you? All right, here are the languages that didn't make the top ten list in order of popularity:

German, Japanese, Urdu, Punjabi, Korean, Telugu, Tamil, Marathi, Cantonese, Wu, Vietnamese, Javanese, Italian, Turkish, Tagalog, and Thai.

If we agree with Marshall McCluan that the "...medium is the message" then it's understandable why there is so much misunderstanding in the world today. 6,500 reasons. Think about the problems we have communicating with each other even when we speak the same language. It's a long journey to overcoming the language barrier.

MEDIUM The 9/11 Commission Report

When the message is lost because the medium (radio equipment) is faulty, the results can be disastrous.

According to the final report of the National Commission on Terrorist Attacks Upon the United States, there was a communication melt down on September 11.

Four main factors prevented successful evacuation instructions to be transmitted by FDNY:

The type of radio equipment being used was not strong enough under the circumstances especially in the high rise environment.

Tactical Channel 1 radio frequency was overwhelmed by too many units trying to broadcast all at the same time.

Some firefighters were not on the right channel probably because of the frustration of not being able to gain access to Channel 1.

Some responders did not have any communication equipment at all.

Equipment failure was only partly to blame for the lack of coordination between the First Responder Agencies including FDNY, NYPD and PAPD.

Analysis of what went wrong is just the first necessary step. A commitment to provide the resources needed by our first responders who put their lives on the line is the second step. A good working relationship between all government and civilian organizations involved in emergency response is the third step. And finally, in the words of the commission..." a rededication to preparedness is perhaps the best way to honor the memories of those we lost that day."

The Gaps

If we have all the basic elements including the Sender, the Message, the Medium, and the Receiver, is there communication? Not if the Interpretation is incompatible with the Sender's intent . Miscommunication is usually because of a Gap. It could be a Generation Gap, A Gender Gap, a Cultural Gap or an Organization Gap.

INTERPRETATION Generation Gap

Our godchild, Eva Marie, is a junior in high school. She's a bright, wonderful girl who is so willing to share with her "elderly" godparents the language of hi-tech. She recently invited us to join her Hi5 network which means we can immediately be connected to her circle of friends and tap into the online service to view photos, browse profiles and enter the chat rooms.

Have you ever been in a chat room or communicated with a teenager via text messaging? If so, then perhaps you encountered the typical chat room speak.

BAK (back at the keyboard)
WB (welcome back)
BG (big grin)
JTLYK (just to let you know)
TOY (thinking of you)
WIBNI (wouldn't it be nice if)
PDA (public display of affection)
IWALU (I will always love you)
H&K (hug and kiss)

TTYL (talk to you later)
KIT (keep in touch)
B4N (bye for now)

Every generation has it's own secret code, a method of talking with one another. We're thankful that Eva thought enough of us to invite us into her generation's world. Our suggestion to parents with teens is that it might be wise to learn some of this new language. Even if you aren't fluent in "chat speak", be on the look out for the following:

NP (nosy parents)
P911 (my parents are coming)
PA (parent alert)
PAL (parents are listening)
PANB (parents are nearby)
POS (parent over shoulder)
TAW (teachers are watching)
WTGP (want to go private)

Chances are whoever's using these acronyms might need more watching.

INTERPRETATION Gender Gap

Maybe men are from Mars and women are from Venus. As long as the population keeps growing, there's definitely reason to believe that at least some of the inhabitants from the two planets are getting together on Earth for some R & R.

No one can deny that there are often barriers to communication between the sexes. But sometimes, it's those very differences that give us reason to enjoy one another. We'd like to share with you a memo we recently received from one of our friends which illustrates what happens when men really listen to the semantics involved.

Words Women Use

FINE

This is the word women use to end an argument when they are right and you need to shut up.

FIVE MINUTES

If she is getting dressed, this is half an hour. Five minutes is only five minutes if you have just been given 5 more minutes to watch the game before helping around the house.

NOTHING

This is the calm before the storm. This means "something," and you should be on your toes. Arguments that begin with 'Nothing' usually end in "Fine".

GO AHEAD

This is a dare, not permission. Don't do it.

LOUD SIGH

This is not actually a word, but is a nonverbal statement often misunderstood by men. A "Loud Sigh" means she thinks you are an idiot and wonders why she is wasting her time standing here and arguing with you over "Nothing".

THAT'S OKAY

This is one of the most dangerous statements that a woman can make to a man. "That's Okay" means that she wants to think long and hard before deciding how and when you will pay for your mistake.

THANKS

A woman is thanking you. Do not question it or faint. Just say you're welcome.

INTERPRETATION The Cultural Gap

Until we really get to know people from a foreign country we tend to stereotype everyone from a geographical area in the same negative light.

For example....

Americans see the French as*4...

- Arrogant and conceited
- Distant and difficult to meet
- People who are rude, anti-American and ungrateful
- People who live in a bureaucratic Socialist system
- People who are totally dependent on the State
- People who do not work
- People who do not use soap
- People who don't want to speak English

The French see Americans as...

- Arrogant, moralizing and right wing religious
- People who are naïve
- People who have no history
- People who only think about money
- People who have free access to guns and use them to shoot each other

- People who do not understand other nations
- People who do not take criticism
- People whose media never address international issues

Things are not always what they seem. A year ago our son Jimmy and his wife Nina invited us to travel to Europe with them – Rome, Florence, Venice and, of course, Paris. In comparison to the Italians, we thought the French people we met were distant and cool but we wanted to reserve judgment.

The day after we arrived in France, we set out early in the morning to tour the Louvre Museum and the Eiffle Tower. When we returned to the Charles-de-Gaulle Hotel late that afternoon around 5:00 P.M., the concierge came up to my wife and informed her that security had her purse. My wife said she didn't lose her purse and that she left it in our room. The gentleman in the tuxedo persisted. "No, no, Madame," he continued. He then explained that the taxi driver who picked us up in the morning and dropped us at the Louvre found her purse in the cab and brought it back to the hotel. Jimmy, whose alter ego is Alex P. Keaton, was about to take over the conversation until he saw the look on his mother's face (bright red) indicating that she realized the concierge was right. We had been warning her over and over to be careful about losing her purse so she quit bringing it with her. This time she did bring it along and forgot it. We thanked the concierge, the security personnel and made arrangements to have the same taxi driver pick us up the next day to thank him personally.

So what about the French? There are some who are cool, arrogant and rude. There are others who are honest, hard working and decent. Just like back home.

INTERPRETATION The Organization Gap

Every organization or association has it's idiosyncrasies. The most obvious is their form of communication. There seems to be an unwritten rule that when communicating with another member, one has to use the same jargon so that non-members won't have any idea what you're talking about. Some examples that come to mind are:

- A doctor discussing your condition with a colleague using words that sound like ancient Greek... "anaphylactic...angina pectoris"...
- A lawyer talking about a case in legalese with your lawyer. Both pretend you're the invisible man... "But, clearly, the alleged altercation..."
- A dentist talking on the phone about the proper surgical approach with the orthodontic specialist ...while you're still in the chair waiting for the novocaine to kick in...
- The pharmacist squinting like she can't read the prescription your doctor sent along...
- Two accountants punching their calculators feverishly on April 14, no words need to be spoken...
- A policeman who stopped you for not wearing a seat belt is talking to someone on the car radio. The only thing you can make out above the static are the words "perp" and "vehicular"...

Of course, the problem is even greater when someone from one organization tries to communicate directly with a non-member. Hopefully, the organization person will recognize that he/she needs to shift gears and use standard language when dealing with the general public.

It could be as simple as just saying 6:00 P.M instead of 1800 hours. Or as uncomplicated as substituting "the person in charge of this study..." for "hazard mitigation assessment and IC effectiveness committee Chair" ...

If we can remember to close the gaps, we might be wonderfully surprised at the decrease in miscommunication and the increase of understanding with one another. It's certainly worth the effort, don't you agree?

THE FOUR CATEGORIES

1. Enduring communication: Timeless Principles

Universal truths are passed down to us from the scriptures of Christianity, Judaism, Islam, Hinduism, Confucianism, Buddhism, and others, including the wisdom of the ages taught by the great philosophers...

Examples:

The Golden Rule
Love Thy Neighbor
Honor Thy Father and Mother
Judge Not
Speak Truth
Seek and Ye Shall Find
Do not harm anything

2. Reserved Communication: Time Delayed Lessons

We listen and go through the motions but we don't really understand the importance of these lessons until that "Aha" moment when we are forced to apply what we have learned.

Examples:

Parents serving as role models on how to raise a family...

Coach talking about the importance of physical fitness...

Teacher introducing life skills in finance, cooking, carpentry...

Nurse explaining the harmful effect of smoking...

Instructor demonstrating CPR and First Aid...

Manager explaining how to handle difficult customers...

Police promoting the value of Neighborhood Watch...

3. Ordinary Communication: Real Time News

We are bombarded by so much data every day that we are in danger of overload –and not enough time. The solution is to focus on our priorities and reserve those special moments with our families and loved ones.

Examples:

Conversations with spouse, children and pets

Daily newspapers, magazines, talk radio, television news

Meetings with co-workers, colleagues and associates

Greetings to customers, clients, acquaintances, strangers

Conferences with the boss, kid's teachers, neighbors

Social activities with relatives and friends

Gatherings at church, community associations, political groups

4. Emergency Communication: Time Sensitive Warnings

Under emergency conditions, time is critical and any delay may result in loss of life. It is imperative that action be taken as soon as possible. The message must be clear, simple and direct. Use any and all resources available. Anyone who can help or bring help is the intended receiver. Leave no room for error in interpretation.

Examples:

An elderly man suffers a heart attack while dining in a restaurant

You are on your cell phone when you see a pedestrian hit

Code Adam is announced over the PA system at Wal*Mart

A woman is screaming "Help" as she's forced into a car

Smoke is coming out of the second story of a neighbor's house

THE KEYS TO UNDERSTANDING

Deciphering two unbroken codes can be your key to a fortune in gold treasure buried in the mountains of Virginia since 1819. Known as the "**Beale Treasure**," the story has been told on the History Channel, Travel Channel, BBC, Fox TV and hundreds of articles.

One version of the story recounts how Thomas Jefferson Beale and his companions accidentally stumbled upon a thick vein of gold while looking for a pass through the Rocky Mountains. Since all of the men were good friends, they decided to split the fortune evenly and made plans to bury the treasure in a safe location back in Virginia. Over a four year period, the group accumulated an impressive amount of gold and silver which was transported and buried in the same unmarked location.

Beale and several of his partners then devised a trio of complex codes.

Code No. 1 describes the location of the treasure.

Code No. 2 describes the contents hidden in the secret cove.

Code No. 3 lists the names of the 30 men who were in the group.

The codes were then placed in a metal strongbox and given

to Robert Morris, the trustworthy innkeeper at Lynchburg for safe keeping. His instructions were to open the box if no one returned for it within10 years.

Morris forgot about the box and it was only after 23 years had elapsed that he came upon the locked box where he had stored it long ago. When he broke open the box, he found a letter addressed to him.

In the letter, Beale recounted in elaborate detail the expedition to Colorado, the discovery of the fortune in gold and silver and the secret trips to the Blue Ridge Mountains of Virginia to bury the treasure. In conclusion, Beale instructed Morris to locate the treasure and divide it up into 31 equal shares – one for each of the names on the list and one part for himself. He only had to use the key which was to be mailed to him from a friend in St. Louis. With the key, he would be able to decode the three sheets of numbers ranging from one to four digits arranged randomly.

The only problem was that Morris never received the letter containing the key nor did he ever hear from Beale again. Morris later shared the letter and the codes to a close friend, James Ward, who surprisingly was able to break Code No. 2 simply by accident. He discovered that the code was based on the Declaration of Independence. It was easy then to decipher the numbers and put together the following hidden message from Beale:

"I have deposited in the county of Bedford about four miles from Buford's Inn in an excavation or vault six feet below the surface of the ground the following articles belonging to the parties whose names are given in number three herewith. The first deposit was ten hundred and fourteen pounds of gold and thirty eight hundred pounds of silver. This was deposited November, 1819. The second deposit was made Decem-

ber, 1821 and consisted of twelve hundred and seven pounds of gold and twelve hundred and eighty-eight pounds of silver.

The above is packed securely in iron pots with iron covers. The vault is lined with stones and the vessels lie on solid rock and are covered with other stones. Paper number one describes the exact location of the vault so no difficulty will be had in finding it.." *5

Energized by this discovery, Ward eagerly worked day and night to solve the riddle of Code No. 1 which contained 520 numbers. But there must have been separate keys to each because the Declaration of Independence which he used to crack Code No. 2 did not fit the pattern. Frustrated and discouraged, Ward finally gave up. To this day, Codes No. 1 and No. 3 remain unbroken and the secret to Beale's Treasure is still a mystery.

PASSWORDS

Whether you believe in the hidden treasure or you are skeptical that it was all just a hoax, you must agree that sometimes trying to understand people is as frustrating as solving a puzzle without the clues. The good news is that we offer the following passwords as your keys to discovering a gift more precious than gold – knowledge and understanding.

Listener: Be an active listener. In order to get the whole message, we have to pay attention to facial expressions, body language, tone of voice and the pacing and speed of the person we are conversing with at the time.

Openness: Open your mind and your heart to whomever is speaking. Do not just hear the words. Sometimes what is not being said is even more important than what is being verbalized.

Versatility: Be flexible and allow for compromise. Give the person the benefit of the doubt before interpreting his/her message in a negative way.

"What's important to the other person must be as important to you as the other person is to you."*6

Empathy: Trade places in your mind's eye for just a minute and see the situation from the other side of the table. "Walk a mile in my shoes."

WATCH YOUR BACK

"Gladys, what siren are you talking about?
I don't hear anything."

DRAW: Bob with his back toward the ocean. A large wave (Tsumani) is behind him.

PART 2

Multiple Dangers/ Multiple Risks

"One clear lesson of September 11 is that individual civilians need to take responsibility for maximizing the probability that they will survive, should disaster strike."

——— *The 9/11 Commission Report*

Directory of Multiple Dangers/Multiple Risks

MEDICAL EMERGENCIES	Heart Attack	77
	Stroke	78
	Seizure	78
	External - Internal Bleeding	79
	Allergic Reaction	80
	Poisoning	80
	Wounds	81
	Burns	81
	Temperature Extremes	82
ACCIDENTS	Motor Vehicle Accidents	85
	Preventable	87
	Not Preventable	89
MISSING PERSONS	Runaways	92
	Wanderers	93
	Abduction	95
	Stranded	99
VICTIMS	Victims of the Wild	101
	Victims of Crime	111
FIRE	What, Where, When, Why, How	122
NATURAL DISASTERS	Earthquakes	129
	Floods	132
	Hurricanes	135
	Landslides	138
	Tornadoes	141
	Tsunamis	144
	Volcanic Eruptions	147
	Winter Storms	150
TERRORISM	Chemical Weapons	157
	Biological Weapons	157
	Radiological Weapons	157

According to researchers at the Mayo Clinic in Rochester, Minnesota, optimistic people had 50% less risk of early death compared to pessimistic people. Since we consider ourselves optimistic personalities, we decided to find out what other factors increase the chances of a longer life. We found the following list from FORBES magazine.

TEN WAYS TO LIVE LONGER:

1. Don't oversleep.
2. Be Optimistic.
3. Have more sex.
4. Get a pet.
5. Get a Vap (The new cholesterol test).
6. Be rich.
7. Stop smoking.
8. Chill out.
9. Eat your antioxidants.
10. Marry well.

For the most part, we both agree with Forbes' top ten list with the following exceptions:

- We would rearrange the priority, admitting that our own priorities would not necessarily be the same.
- We confess that we didn't really know what a Vap is or what benefits antioxidants presented, but now that we've done the research we can understand why they were included.
- One of the most important factors not included in the list is the ability of an individual in certain situations to control the causes of death.

Compare the list of Leading Causes of Death in the U.S. with the top ten Behavioral Causes:

LEADING CAUSES OF DEATH IN THE U.S.

1. Heart attack
2. Cancer
3. Stroke
4. Respiratory Disease
5. Accidents
6. Diabetes
7. Flu & Pneumonia
8. Alzheimer's senility
9. Kidney disease
10. Infections
11. Suicide
12. Liver disease
13. High blood pressure
14. Homicide
15. Other

TOP TEN BEHAVIORAL CAUSES OF DEATH

1. Tobacco
2. Diet/Activity Patterns
3. Alcohol
4. Microbial agents
5. Toxic Agents
6. Firearms
7. Sexual behavior
8. Motor Vehicles
9. Illicit use of drugs
10. Accidents

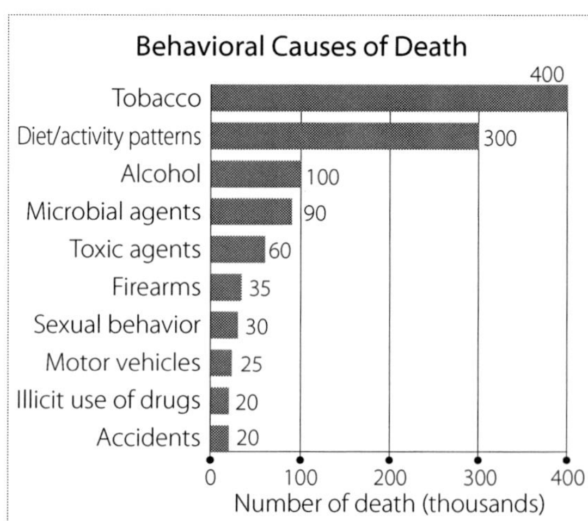

Source: 1993 study by Carter Center

The lesson, it seems, as regards our behavior, is threefold:

1. If we can control what we inhale, eat and drink...
2. If we can control whom and what we play with...
3. If we can control our impulses to be careless and foolhardy...

Then perhaps we will live long enough to see our children and grand-children grow up and marry well, get a pet and be successful as well.

We have divided **Multiple Dangers/Multiple Risks** into seven major categories:

- Medical Emergencies*[1]
- Accidents
- Missing Persons
- Victims
- Fire
- Natural Disasters
- Terrorism

Our intent is to present the basics in substance for each category and to encourage a comprehensive approach to all dangers/risks by following the formula we have presented in our introduction- the POO Factor and the Power of 3.

If you ask the question "What three things should I do ...? for each of the following:

- Immediate Danger
- Imminent Danger
- Continued Risks

If you proceed to do the things necessary for:

- Immediate action
- Crucial Follow Up

- Long Term Goals

Then we believe you will have gone a long way in making sure your family is ready in any emergency.

MULTIPLE DANGERS / MULTIPLE RISKS

Medical Emergencies

"A merry heart does good like medicine."

——— *Proverbs 17:22*

Imagine the busiest highway at peak travel time (4:30 p.m. on a Friday, the beginning of a three day holiday) in your city. Bumper to bumper as far as the eye can see, isn't that right? Now picture your house. O.K. Now imagine an ambulance trying to negotiate the traffic from the hospital through all that rush traffic to your house. How long do you think it would take? Too long if there was an emergency. That's why it's so important to have ordinary people with basic skills who can be a bridge to the medical professionals – someone who can fill in temporarily until the emergency personnel arrives. If you could do something to help your family and your neighbor, wouldn't you? Basic knowledge, basic skills in First Aid, CPR (cardiopulmonary resuscitation) and the use of the AED (automatic external defibrillator), can help save a life. Why not you? Why not now?

HEART ATTACK

— What, Who, How —

If one or more of the coronary arteries which supply blood flow to the heart muscles is blocked, then part of the heart muscle or tissue dies. This is known simply as a **heart attack**. The technical medical terms are coronary thrombosis or myocardial infarction.

Symptoms: Tightness in chest; pain behind breastbone

Severe indigestion, nausea
Pain in the arm, shoulder and back
Short, fast breathing
Pale, bluish color; sweating

Treatment: Call 911 – don't wait
Keep victim calm ; do not leave victim alone
Do not swallow anything except prescribed heart medication
Perform CPR if needed
Use AED if available

STROKE

— What, Who, How —

A sudden and dramatic drop in blood pressure causes a life threatening condition known as **circulatory shock**.

Symptoms: Sudden numbness in face, limbs, especially on one side
Trouble speaking or understanding
Blurred vision, dizziness
Loss of balance and coordination
Severe headache

Treatment: Call 911 immediately
Stay with victim until help arrives

SEIZURE

— What, Who, How —

When there is an instability of electrical activity in the brain, a partial (affecting only a portion of the brain) or generalized (affecting the whole brain) seizure can occur.

Symptoms: Fainting or complete blackout
Drooling
Muscle spasms, twitching and thrashing
Loss of bladder control

Treatment: Call 911
Do not put anything in victim's mouth
Do not restrain the victim
Do not move the victim unless necessary
If victim is convulsing with high fever, cool body with wet compresses – do not put victim in tub

Take Note: How long the seizures last; If there are multiple seizures
If the victim can be awakened between seizures
If victim has diabetes, high blood pressure or alcoholism

EXTERNAL-INTERNAL BLEEDING

— What, Who, How —

An injured person may have internal bleeding caused by a fall, a beating such as a kick or punch, or an ulcer. A gunshot or stab wound which causes bleeding externally could also be the cause for internal injuries requiring immediate medical attention.

Symptoms: Vomiting, coughing up blood
Bleeding from the rectum or vagina
Abdominal swelling
Painful, swollen , bruised areas
Cold, clammy and pale skin

Treatment: Keep victim warm and relaxed
Seek medical attention immediately

ALLERGIC REACTION

— What, Who, How —

Someone who comes into contact with a trigger substance can have a severe allergic reaction which leads to anaphylactic shock. This is a life-threatening condition and must be treated right away.

Symptoms: Swelling of tongue and lips
Difficulty breathing
Nausea, vomiting
Hives and itching

Treatment: Call for medical assistance
Keep victim warm and calm
Moisten lips but no food or liquid
Lay victim down and elevate legs if there are no injuries

POISONING

— What, Who, How, Why —

Any substance, taken in sufficient quantities, that can cause either temporary or permanent damage to the body can be defined as a toxin or poison. Poisons can be swallowed, inhaled or introduced to the body through the skin. Poisoning can occur accidentally or intentionally.

Categories of Poisons: Drugs and alcohol overdose
Household chemicals
Medicines and pesticides
Bites and stings
Plants and fungi
Bacterial and viral toxins in food
Carbon monoxide gas

Treatment: Call Poison Control Center and hospital physician
Take victim to emergency room or call 911
Identify the source of the poisoning if possible

WOUNDS

— What, Who, How —

Loss of too much blood can cause shock and is life threatening. The primary goal is to control the bleeding. Do NOT remove any penetrating object which has caused the wound. Apply enough bandages to support the object and keep it from moving.

Symptom: Severe bleeding
If blood is bright red, there may be an artery involved

Treatment: Call 911 and advise status
Have victim lie down and keep calm
Apply pressure over the wound with a clean, absorbent cloth
If bandage becomes soaked, apply another
If bleeding cannot be controlled, apply pressure at a point above the wound
Control bleeding until ambulance arrives

SEVERE BURNS

— What, Who, How —

If a person suffers severe burns to the face, hands, feet, mouth or groin, seek medical attention as soon as possible.

First-degree burn: affecting only the top layers of skin

Symptom: Redness and pain

Treatment: Run under cold water over affected area
Keep clean to prevent infection

Second-degree burn: affecting the underlying and outer layers of skin
Symptom: Redness, pain, blisters
Treatment: Run cold water over affected area
Cover the burn with clean gauze loosely
If blisters open, keep covered with clean dressing
If large area is affected, go to clinic or hospital

Third-degree burn: all the layers of skin and even the nerves are affected
Symptom: Skin will look black or white
Underlying skin or tissue is exposed
Extremely painful or numb, if nerves are damaged
Treatment: Call 911
Cover burned area with sterile gauze, do not use tape

TEMPERATURE EXTREMES

— What, Who, How —

If the body is not given enough time to adjust to sudden extreme temperature changes, the result may be a serious life-threatening condition.

Heat Stroke: This occurs usually in very hot environments or because of a fever. A very serious condition, heatstroke is often fatal if not treated immediately.
Symptoms: Red, hot, dry skin; sweating has stopped
Weak pulse, shallow breathing

Confused, delirious, irrational
Muscle cramps, convulsions

Treatment: Seek medical help immediately
Go to a cool, air conditioned place
Cool down skin with cold water or wet cloths
Take sips of water or sports drink

Hypothermia: Hypothermia can occur if the body temperature drops below 95 degrees F. Temperatures do not have to be freezing. It can happen if a person is immersed in cold water ; is exposed outdoors in wet and windy weather; or lives in a poorly heated home. The elderly, infants, and the chronically ill are highly susceptible to hypothermia.

Symptoms: Shivering, numbness, rigid muscles
Slurred speech
Slow pulse, glassy stare
Not fully conscious

Treatment: Call for medical assistance
Remove wet clothing; keep warm
Drink something warm without alcohol or caffeine
Do not immerse in water

COMMON SENSE STRATEGIES *» Medical Emergencies*

- Learn basic First Aid techniques
- Learn how to perform CPR (cardiopulmonary resuscitation)
- Learn how to use the AED (automatic external defibrillator)

Note: The American Red Cross offers certification in all of the skills listed. Check with the American Heart Association for their program offerings. Also, many fire stations offer free non-certified classes to the public.

Accidents

> "The ideal man bears the accidents of life with dignity and grace, making the best of the circumstances." ——— *Aristotle (384-322 BC)*

An accident is usually defined as "a happening that is not planned or expected." The difficulty inherent in this definition is that people often tend to use it as an excuse for not accepting any kind of responsibility. "It's not my fault, because it was an accident." Sometimes, they are right and that's where the difficulty lies. How do we make the distinction?

Following the guidelines outlined in our ABC's (Awareness – Behavior – Control), we can now characterize accidents as falling into two different categories: preventable and not preventable. Let's think about that and decide whether you agree or not after we look at some statistics.

LEADING CAUSES OF ACCIDENTAL DEATHS:

1. Motor Vehicle Accident (MVA)
   Includes car wrecks and pedestrians
2. Falls
   Especially off ladders, down stairs, through windows
3. Poison (solid or liquid)
   Medicines, drugs, mushrooms, etc.
4. Drowning
   Ocean, boat accidents, pools, in the bath

5. Fires

   Smoke inhalation and burns; 40% die in their sleep
6. Suffocation

   Unable to breathe because of blockage by food etc.
7. Firearms

   75% young males (14 – 25) shoot self or others
8. Poisoning (gas)

   Mostly carbon monoxide from car, heater, appliances
9. Medical & Surgical Complications

   Problems in the hospital – no, this is not a mistake
10. Machinery

    Usually on the farm, factory, construction sites

Are you surprised by any of the top ten categories ? We can hear some of you gasp at NO. 9. O.K. Take a deep breath and let's continue. Since MVA's (motor vehicle accidents) are NO. 1, let's do a little more analysis.

MOTOR VEHICLE ACCIDENTS

Serious traffic violations were committed in 55% of all fatal motor vehicle accident:

- Speeding
- Failure to yield right-of-way
- Crossed the center line
- Passed stop sign
- Improper passing
- Disregard traffic signal
- Followed too closely

Source: National Vital Statistics Report (September 2002)

The reason we cite these violations, is to reinforce the obvious link between dangerous erratic driving patterns and the two principal causes:

Driving under the influence of alcohol is the primary cause of death in automobile accidents.

Driver fatigue is the second most frequent cause of fatality in motor vehicle accidents.

Impact of Alcohol Abuse:

There is no question that excessive drinking impairs judgment and ability.

More than a third of pedestrians killed in 1992 in motor vehicle accidents were intoxicated. Moreover, between 1975-1995, alcohol intoxication was responsible for :

32% of fatal falls

42% of fatal fires

34% of fatal drownings

29% of fatal poisonings

32% of homicide victims

23% of suicide victims

Source: Annals of Emergency Medicine 33 (1999)

To clarify the top ten list, the rankings change depending on whether the accident happened at home or at work.

Home Accident

- Falls
- Poison (solid/liquid)
- Fire, burns
- Obstructed airway
- Suffocation

- Firearms
- Poison (gas)

Workplace Fatalities

- Transportation
- Violence
- Machinery, falling objects
- Environmental contamination
- Falls
- Fires
- Explosions

Now that we've reviewed the data, let's revisit the question of separating accidents into the two categories: preventable and not preventable.

Preventable:

Now that we know that driver fatigue is the second leading cause of death in motor vehicle accidents, wouldn't you take a few extra minutes and stop for that hot black cup of coffee? Or at least pull over at the next rest stop and walk around the car to take a breath of fresh air? Who knows how many deaths could be prevented by that one simple act by people who are aware.

In 1979, the Oxford Hills area in the state of Maine suffered the traumatic loss of seven alcohol related teen deaths during the graduation season. As a result of this tragedy, the community created an alternative to the traditional graduation night celebration which too often resulted in excess drinking and driving. This novel idea was named "Project Graduation." Within the relatively short time of seven years, Project Graduation activities took hold from coast to coast and is being celebrated today in all fifty states.

In 1996 when Kathie Wells, the Project Grad chairwoman of Kalaheo High School in Hawaii, was seeking donations to help boost attendance at the school's graduation party, she found an angel waiting with a gull-winged Bricklin. Kathie asked Mike Mckenna, honored as National Auto Dealer of the Year and known for his support of local schools, for a $100 donation. Instead, he gave her a free car as a door prize which resulted in attendance at Project Grad increasing from 35% to 97%. That was only the beginning of his generosity. Mckenna has given away more than 25 cars to date.

How many accidents do you think were prevented over the years throughout the entire country because of one community's great idea and the generosity of angels like McKenna?

Mother's Against Drunk Driving (MADD) just celebrated their 25th anniversary. Since the founding of the organization, alcohol-related motor vehicle fatalities have declined 43 percent. MADD's mission is "to stop drunk driving, support victims of this violent crime and prevent underage drinking." They believe that drunk driving is 100% preventable.

If you don't mind, could we talk about NO. 9 again. The bad news is that the Centers for Disease Control and Prevention estimate that more than two million patients suffer hospital-acquired infections each year. The good news is that the Institute for Healthcare Improvement with other leading American health care organizations has launched the 100,000 Lives Campaign. The campaign aims to enlist thousands of hospitals across the country to commit to six basic changes in care which have been proven to prevent avoidable deaths.

- Deploy Rapid Response Teams – at first sign of patient decline
- Deliver Reliable. Evidence-based Care for Acute Myocardial Infarction - To prevent deaths from heart attack
- Prevent Adverse Drug Events
- Prevent Central Line Infections
- Prevent Surgical Site Infections – reliable delivery of perioperative care
- Prevent Ventilator-Associated Pneumonia

Sorry, we know we promised to use only uncomplicated language. You'll be pleased to know, however, that the campaign is working. Over a thousand hospitals have committed to the program. Morale at these hospitals is way up because they know that their efforts are drastically curbing in-house infections. For example, at Newark Beth Israel Medical Center in New Jersey, not one case of pneumonia in a patient on a ventilator was reported for over a year. Usually, one case every week is common for ventilator patients and usually the survival rate is not good. We applaud the 100,000 Lives Campaign for their successful beginning.

Not Preventable:

Imagine that you are asleep when suddenly you are awaken by a loud crash and the whole house shakes. You think it's an earthquake so you quickly stumble out of bed and rush to your children's bedroom to make sure they are all right. You stare in disbelief at the large elm tree from your back yard now protruding through the roof of the house. Your sleeping daughter is buried underneath one of the large limbs. Quickly you call 911 and the ambulance rushes her to the hospital in a coma...

Imagine that the heavy rains have caused a landslide and the traffic is stalled because of falling trees and rocks blocking the road. You watch in horror as several large boulders come crashing down the side of the mountain and hit one of the road crew clearing the debris. He is knocked unconscious.

These are just two true-life examples of accidents which are not preventable. We're sure that you have more incidents which you have experienced in your lifetime. Bad things do happen to good people. Sometimes there is the chance that you are just in the wrong place at the wrong time. But we believe that more often we are able to anticipate a crisis and be proactive in the prevention of accidents overall. What three things can you do to prevent an accident from happening in the top ten fatal categories? Here are some suggestions.

COMMON SENSE STRATEGIES » *Accidents*

- Don't drink and drive. Don't doze and drive. Use Designated Drivers.
- Safety proof your house for your elderly parents and children. Use grab bars in tubs, railings on stairways, skid resistant pads for carpets.
- Nobody climbs a ladder without a backup.
- Keep all medicines, chemicals and poisons stored in safe places where children can't get to them.

- Supervise your children at all times when near water. Teach them to swim.
- Install smoke detectors. Know how to use fire extinguishers, Have two exits for every room. Practice escape plans.
- Choose food and snacks that are age appropriate for your children. Learn CPR techniques.
- Parents must keep all firearms unloaded and locked up. Gun owners should have safety locks on all firearms. Children should be taught to tell an adult if they find a gun and not to touch it.
- Fix broken appliances, gas heaters, or defective car mufflers as soon as possible. Do not run any type of engine in an enclosed area. Remember carbon monoxide is odorless.
- Ask your hospital if they are members of the *100,000 Lives Campaign*.
- Always stay focused and alert when operating life threatening machinery.
- Use only hands free cell phones while driving.

Remember! Distraction can kill.

Missing Persons

> "People should learn to see and so avoid all danger. Just as a wise man keeps away from mad dogs, so one should not make friends with evil men."
>
> ——— *Buddha*

THE FOUR CATEGORIES:

1. Runaways
2. Wanderers
3. Abduction: Family Abduction; Stranger Abduction
4. Stranded

CATEGORY **1**

RUNAWAYS

— Who, Why, When, Where —

The difference between a runaway and a missing child is intention. A missing child may be injured, lost, or held against their will by a family member or a stranger. A runaway purposely leaves home.

Motivations of a Runaway:*2

- To escape a painful or difficult environment
- To avoid a punishment or other responsibility
- To seek out other people who are supportive
- To avoid the loss of relationships or friendships
- To pursue activities which are important
- To change their present situation

COMMON SENSE STRATEGIES
» Runaways

- Try to establish honest communication and seek to build trust.
- Establish a crisis intervention plan for your teenager.
- Attend a conflict resolution program as a family.
- If alcohol or drug abuse is a contributing factor, contact a qualified alcohol and drug counselor.
- Ask advice from a qualified mental health professional for any emotional problems associated with violent or suicidal behavior.
- For any medical condition requiring treatment or medication, ask for a second opinion.

WANDERERS
Lost Alzheimer's Patients
— Who, Why, When, Where —

CATEGORY **2**

One of the dangerous effects of Alzheimer's Disease is a common behavior referred to as "Wandering" during the night or day. Experts project that 60% to 70% of all Alzheimer's patients will wander at some point. The frightening aspect is that 50% of the patients who get lost will probably die if they are not found within 24 hours.

Causes of Wandering:

- Medication side effects
- Memory loss
- Confusion
- Stress
- Loss of language
- Feeling of Obligation: work, family etc.

All family members must give their full cooperation to the designated care givers. It takes just one brief moment for an Alzheimer's patient to walk out unnoticed. Some suggested ways to prevent wandering:

- Install alarm systems, monitoring devices and locks.
- Use warning signs on doors such as "Do Not Enter."
- Identify the times the patient is most likely to wander in order to keep him/her busy.
- Encourage visitors and arrange fun activities.
- Reinforce and encourage the patient to relieve the stress of confusion.
- Have your emergency plan ready.

Safe Return Program

The National Alzheimer's Association partnered with the U.S. Department of Justice to create **Safe Return**. Since 1994, the program is credited with aiding the return of 7,400 lost patients.

Here's how the program works:

- Contact the Alzheimer's Association at 1-888-572-8566.
- For a one-time fee of $40.00 and a $5.00 charge for caregiver jewelry, the patient receives
    - Identification materials including a wallet card, bracelet, clothing labels, lapel pin and bag tags. These materials indicate the patient is memory impaired.
    - The patient's name, photo, identifying characteristics and emergency contacts are placed in a national data base.
    - An emergency toll free number is manned 24 hours a day to report a lost and/or found patient.

Digital Angel

Locator devices can be used to pinpoint the location of lost Alzheimer's patients. The Digital Angel can be mounted in a car, used as a pager or as a wristwatch. A parent or an authorized person can track the location of the wearer at all times. An alert can be sent to a cell phone, computer, personal digital assistant or text pager whenever the wearer presses a "panic" button, wanders out of a pre-determined boundary or other sensors indicate an emergency situation.

Emergency Procedures

In the event a patient is reported missing, call 911 and follow prepared Search and Rescue plan immediately. Do not hesitate.

CATEGORY **3**

ABDUCTION

Family Abduction

— Who, Why, When, Where —

Parents need to worry more about other family members and acquaintances rather than strangers. According to a report by the Office of Juvenile Justice and Delinquency Programs, the most common kidnapper is a member of the child's family.

Family :	49%
Acquaintances:	27%
Strangers:	24%

Five Things You Might Not Know About Family Abductions

1. Every year in the U.S., over 203,000 children are abducted by a family member, usually one of the parents.
2. The motive is usually revenge against the other parent.
3. More than 50% of the parents involved in an abduction has

a criminal record, a history of substance abuse and a previous background of violent behavior.

4. Studies show that abducted children in these situations undergo severe psychological and emotional problems.
5. A majority of law enforcement agencies in the country do not have a policy on how to treat a family abduction.

To date, 38 states have passed the Uniform Child Custody Jurisdiction Enforcement Act, a law which gives authorities more leverage to prevent family abductions.

Source: StopFamilyAbductionsNow.org –a campaign by the Polly Klaas Foundation

COMMON SENSE STRATEGIES » *Family Abduction*

- Teach your children the skills to keep them safe.
- Update your Child Identification file with photographs, fingerprints, Physical descriptions, dental records, DNA samplings.
- Make time to get to know your child's friends and their families.
- Consider carefully the people you let into your child's life on a regular basis.
- Take part in your community's neighborhood watch program.
- Help pass the Family Abduction Prevention Act of 2004.

ABDUCTION
Stranger Abduction
— Who, Why, When, Where —

CATEGORY **3**

A parent's worst nightmare is the thought of their child being kidnapped by a stranger. The truth is that stranger abduction is the least common form of abduction, although it is the most highly publicized. Still, the danger does exist and it can happen to anyone.

No parent wants their child to be included in that small percentage of missing children snatched by a stranger.

It's important that we recognize that the typical human predator will probably not look like the evil monster we envision. Chances are the stranger will look like any other normal person.

The stranger abductor will be one of two types:
The Power Predator : an abductor who uses force
The Persuasion Predator: one who lures a child in various ways

Both types of predators are looking for **opportunity** and **access**. If we find the way to deny the predatory criminal what he's looking for – a child who is alone or one who is vulnerable - then we can protect our children.

The Challenge

The best way to help our children and to close the door on any would be predator is to have an open communication with our entire family. Everyone in the family should know the risks involved and the reasons for implementing your family plan.

Unfortunately, many parents are in denial so they rationalize and make excuses as to why it isn't necessary. Some of the reasons they refuse to use the most effective tool they have are because:
They don't want to frighten the child
They don't like to talk about such things
They don't know what to do

Parents will do anything to protect their children. Realizing that we cannot always be there in person, it is important to teach our children rules on how to stay safe. We need to teach them to recognize dangerous situations, to be, street smart. We need to give them the tools they need to protect themselves when necessary. Just as we teach them safety guidelines about swimming in the ocean or about safe driving skills or about dangerous drugs, we should also teach them how to avoid abductions.

Some Ways to Protect Your Child

- "No, Go, Tell" Explain to your children that there are situations if they feel uncomfortable with anyone or scared, to say no, then go tell someone that's on the list you've prepared in advance (neighbor, teacher, police...) or go to the first lady they see.
- Create a secret password or secret code. Under no circumstances should the child go with anyone, enter someone's car, go into anyone's house unless they have your permission or the person knows the password. See Form F14.
- Encourage your children to follow the buddy system.
- Establish a plan of what to do in various scenarios such as being separated at the shopping mall or at the movie theater.
- Teach them how to use the telephone, dial 911, how to use a cell phone, talk to an operator, how to make a long distance call, and how to use a pay phone
- Make sure your young children know their full names, the phone number including area code, address and city where they live.
- Encourage your child to enroll in self-defense classes which will help build their confidence and their skills.

Source: Office of Juvenile Justice and Delinquency Program

STRANDED

— Who, Why, When, Where —

CATEGORY **4**

Some of the reasons people get lost and stranded include the following:

- Weather related: caught in a snow storm
- Injured: broken ankle on a hike
- Equipment failure: snowmobile can't start
- Trapped: avalanche blocks only exit
- Abandoned: left behind on an adventure tour unintentionally
- Disoriented: wandered off marked trails

Ideally, the person has told someone where he was going; he has with him some of the essentials (map and compass, flashlight, extra clothes, matches, extra food and water, pocketknife...); he knows how to read a map and compass. Realistically, he hasn't and he doesn't.

COMMON SENSE STRATEGIES *» Stranded*

STOP: Sit. Think. Observe. Plan

- As soon as people realize they are lost, they compound the problem by reacting in panic mode. This means that their brain is running out of sync with their limbs which leads to falling and possible injury. The best way to react to the challenge is to *FOCUS*.

SIGNAL: SOS

- Realize that you won't get help unless someone knows you need help.

- Communicate by using any emergency signaling technique to alert someone to your distress and location. The more creative you are, the more effective your signaling can be.

EMERGENCY SIGNALING METHODS

- Sounds: whistles, gunshot, drums, shouting, anything audible
- Fires and smoke
- Mirrors
- Ground to air messages: SOS
- Change natural landscape and create unnatural patterns.
- Use movement and contrasting color by waving flags to attract attention
- If you anticipate that there's a Search and Rescue party close by, use repeated sounds like whistles to help them zero in on your location.
- During the day, concentrate on using colors and ground to air messages.
- At night, use flashlights, vehicle lights, emergency flashing lights.

Use the Buddy System

It's always safer to have someone with you. Never go alone to the Beach, Video Arcades, Parks, Movie Theaters, Swimming Pools, Public Restrooms, Door to Door for Fundraisers, School Yards or walking anywhere after dark.

Victims

> "Life is like a good book – you can write it as a comedy, a tragedy, or an inspiration."
>
> ——— *A. J. L*

VICTIMS OF THE WILD

It would be impractical for us to cover every possible living thing that can cause harm to human beings. In victims of the wild, we have focused on the more common culprits, the most dangerous, and the most sensationalized. Outdoors in our context could mean the desert in Arizona, the mountains of Montana, the blue waters of the Pacific or the woods in your back yard. Be aware of your surroundings so that you can prevent your family from becoming a victim of the wild.

PLANTS

— What, Where, How —

Poison ivy, western poison oak and poison sumac: All three plants contain a poisonous sap in their leaves, fruit and stems. In the spring and early summer when the leaves are tender, the sap is easily released when bruised. The sap can make contact with the skin directly or through contaminated objects like shoes, clothing or your pet.

Symptoms: Redness of skin and extreme itching
Rash can appear in streaks or patches
Blisters

Treatment: Wash skin immediately with soap and water
Remove and wash clothes and shoes to prevent spreading

The Danger of the Wild

NOTE: ✔ Check the category that applies to you.

	Snakes	Spiders	Killer Bees	Bear	Mountain Lion	Wolf	Alligator	Shark
Alabama								
Alaska								
Arizona								
Arkansas								
California								
Colorado								
Connecticut								
Delaware								
Florida								
Georgia								
Hawaii								
Idaho								
Illinois								
Indiana								
Iowa								
Kansas								
Kentucky								
Louisiana								
Maine								
Maryland								
Massachusetts								
Michigan								
Minnesota								
Mississippi								
Missouri								

Reference Chart

	Snakes	Spiders	Killer Bees	Bear	Mountain Lion	Wolf	Alligator	Shark
Montana								
Nebraska								
Nevada								
New Hampshire								
New Jersey								
New Mexico								
New York								
North Carolina								
North Dakota								
Ohio								
Oklahoma								
Oregon								
Pennsylvania								
Rhode Island								
South Carolina								
South Dakota								
Tennessee								
Texas								
Utah								
Vermont								
Virginia								
Washington								
West Virginia								
Wisconsin								
Wyoming								

Apply anti-itch medication and cool compresses
Call doctor if rash persists or itching can't be controlled
Seek medical attention if there is difficulty breathing

INSECTS, SPIDERS, SEA CREATURES

— What, Where, How —

Bee & Wasps Stings (Fire Ants in South): Shock can result if there is an allergic reaction to the insect sting. Seek medical attention immediately if there is difficulty breathing, coughing and severe headache.

Symptoms: Red, swollen, painful area

Treatment: Remove stinger by scraping skin with edge of plastic card
Wash painful area with soap and water
Apply calamine lotion or 1% hydrocortisone cream
Use ice pack

Spider Bites: Spiders hide in dark areas such as basements, woodpiles, under boxes, furniture, and even in shoes.

A physician should examine spider bites especially by the Black Widow or the Brown Recluse Spider. Bring the dead spider or any remains with you so that the medical professionals can identify it.

Black Widow

Symptoms: Bite looks like small pin prick
Tightness in chest, abdominal pain
Shortness of breath, excessive sweating
Headache, swollen eyelids
Pain in muscles, pain in feet

Brown Recluse Spider

Symptoms: Bite looks like an ulcer with a red ring around it
Stinging sensation
Shock, fever, weakness
Nausea
Joints ache

Treatment: Wash area with soap and water
Apply a cold compress

Jellyfish Sting:

Man-o-war

Warnings of seasonal jelly fish invasions are usually posted on signs at the beach. Check with lifeguards for advice about rip tides, jellyfish and shark sightings.

Symptoms: Severe stinging sensation while swimming in the ocean
Tentacles clinging to skin

Treatment: Seek medical advice if there's an allergic reaction
Gently remove tentacles by using edge of plastic card
Rinse area with water (salt or fresh) or vinegar
Apply hydrocortisone cream on painful area

SNAKES

— What, Where, How —

Rattlesnakes, Coral Snake, Water Moccasin, Scarlet King Snake

Dangerous species of snakes are found in most desert and tropical regions of the continental United States. They like to hide under rocks and underbrush or in grass. It is often difficult to distinguish between dangerous and harmless species. If at all possible, capture or kill snake to show medical professionals. Right now you're probably thinking out loud "Are you crazy?" Well, at the least, try to remember what it looks like so you can describe it. That important information could make a big difference in the treatment recommended.

Symptoms: Sharp pain, swelling
Shortness of breath, dizziness
Vomiting
Blurred vision, weakness

Treatment: Stay calm and get help at hospital or clinic
Remove rings and jewelry from bitten limb
Wash wound with clean water, apply bandage firmly
Do NOT cut the bite or suck out venom
Do not apply ice
Keep the victim calm and immobilize the limb

FOUR FOOTERS

— What, Where, How —

For people on vacation or regular bikers, joggers, hikers, campers and those on picnic outings in the woods, forests or mountains that wildlife inhabit:

Encounters with animals can cause problems for people and pets. Although most animals will not attack humans unprovoked, they will respond aggressively when surprised or cornered and especially when they feel their young are threatened. Attacks by rabid animals can not only cause severe injury but also can spread dangerous infections and rabies. Rabies, if left untreated, will likely cause death.

Bear: Black bears are the most common species found in the U.S., the total population estimated at approximately 300,000 in 41 of the continental states. Despite the fact that they are called Black bears, their coats range from black to cinnamon to blonde. They are relatively small (150 –250 lbs.)compared to their cousins the grizzly bear (300-500 Lbs.)

Bear attacks are uncommon and most encounters occur because you accidentally surprise them; your dog provokes an attack; they are following food smells to you; the bear is used to people and has lost its natural fear.

Avoid an Encounter:	Watch for fresh bear signs and stay away
	Keep your dog on a leash at all times
	Travel in groups
	Make lots of noise
	Never approach a bear
	Be aware of surroundings, use marked trails
Unavoidable contact:	Stay calm, talk in firm tones, avoid eye contact
	Move slowly in the direction you came from
	Do not run – bears can outrun you
	Try to scare it by yelling, or throwing rocks
	Use your bear spray
	Last resort is to rollover and play dead

Mountain Lion: Mountain lions are also known as puma, cougar, panther, painter, and catamount. Reported sightings of cougars have increased dramatically especially in California. One reason is that 80% of all cougar sightings are actually deer, bobcats, dogs and even domestic cats. Another reason is the increase in the human population. Cougars are more closely related to leopards than lions and are experts at hiding while moving about. It is unusual to see the entire cougar in the wild.

Avoid an encounter:	Avoid walking near dense growth or under ledges
	Don't hike, jog, bike alone in lion country
	Don't allow children to play alone outdoors
	Don't feed or leave pets outdoors
	Don't leave any food out that will attract lions
Unavoidable contact:	Don't act afraid; look lion in the eye
	Never, never turn your back or run
	Back away slowly or try to give it a way out
	Make loud noises; make yourself look bigger
	Use mace, pepper spray, rocks, sticks, tools
	Focus on hitting the eyes with fingers or any weapon

Wolf: The gray wolf is on the U.S. Endangered Species List. There are approximately 2500 in Minnesota, 321 in Michigan, 335 in Wisconsin, 81 in Montana, 271 in Yellowstone, 285 in Idaho and about 7500 in Alaska.

Wolves are opportunistic predators. They tend to kill the young, old and sick or injured. When an animal runs away, the wolf's instinct is to give chase. A moose that stands and fights will often force a pack to seek easier quarry.

Attacks by wolves on humans is very rare. The main concern would be an attack by rabid wolves. After the first lethargic phase, an infected animal then dashes around in a frenzy. Rabid wolves can cover miles in a short time and endanger anyone in their path.

Dogs: Any dog, regardless of breed, is only as dangerous as its owner allows it to be. There may be a variety of circumstances which contribute to a dog attack, but the most important factors are:

Owner irresponsibility:	Allowing dogs to roam
	Encouraging dogs to behave aggressively
	Neglecting and abusing dogs
	Allowing access to chained dogs
Parent irresponsibility:	Unsupervised children
	Allowing children to approach unfamiliar dogs
	Not teaching children to respect animals
Avoid an encounter:	Never approach a chained dog
	Stay calm, walk away slowly, don't run

If you are threatened by a dog, pick up a stick and throw it. Pick up stones to throw or pretend to throw. Most dogs will retreat if they think you are armed.

SHARK

— What, Where, How —

The great white, tiger and bull sharks are the largest species that feed on human-size prey. Other species recorded to have attacked people include the hammerhead, short fin mako, oceanic whitetip and some reef sharks like the Caribbean.

Up until 2004, there have been 761 confirmed attacks and 39 deaths in the continental U.S.; 100 confirmed attacks and 15 deaths in Hawaii. The actual number of attacks is hard to rely on because of poor reporting.

The most common type of attack is the "hit and run." The shark bites the person then quickly disappears. We've known three teens who were attacked in this fashion. All three were surfing at the time. Two of the boys lost their legs below the knee. The third teen is celebrated Bethany Hamilton who lost her arm up to the shoulder in the attack. See our interview with her father, Tom Hamilton, and the 10 Tips to Avoid Shark Attack in the Quick Reference section.

ALLIGATOR

— What, Where, How —

Alligators are large reptiles with a scaly dark-colored hide and long muscular tail. They thrive in the swamps and marshes of Florida and are usually found only in fresh water. Crocodiles are so rare and secretive and only nest in the National Wildlife Refuge so the chances are that any giant lizard-shaped reptile you see will be an alligator.

Alligators will eat anything they catch including fish, birds, dead animals, other alligators and human who have the misfortune of being in the water. An alligator can float motionless in the water for long periods of time waiting for its next meal. They are seemingly invisible. Do not go in the water anywhere close to where alligators are known to be lurking.

COMMON SENSE STRATEGIES *» Victims of the Wild*

- Be aware of the types of wild life that inhabit the areas where you live and vacation.
- Dress appropriately for all your activities.
- Bring the right tools and equipment such as bear spray.
- Do all that you can to avoid encounters with dangerous wild life.
- Know the appropriate actions to take in case of a confrontation.
- Learn to recognize symptoms that need medical attention.

VICTIMS OF CRIME

"The most dangerous, cunning and ruthless animal of all, is man." —— *A. J. L*

ANATOMY OF A CRIMINAL MIND

— What, Where, When, Who, Why, How —

The Victim Triangle*3

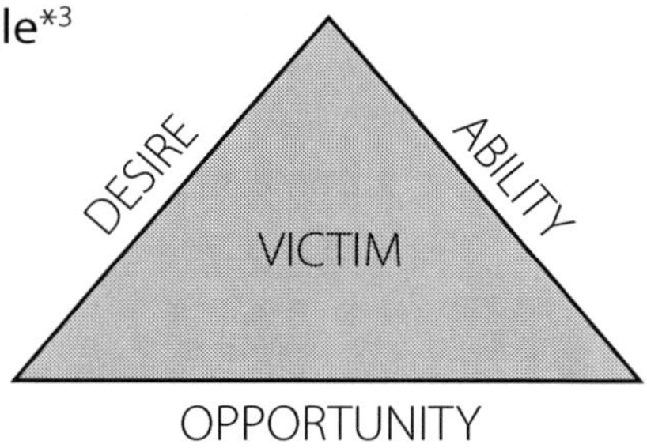

Desire: The criminal has in mind something he/she wants and goes looking for it or sees a person, place or thing that fuels a need.

Ability: The criminal believes he/she has the ability to accomplish the task and escape. There is often a big difference in ability between the practiced criminal and the "first time and only time" offender.

Opportunity: This is the crucial element which can mean the difference between the success of a crime being committed or even an attempt being made. Being AWARE and PREPARED can deny the opportunity to a would be criminal.

Victim: Can be a person, place or thing.

VIOLENT CRIMES

The categories of violent crime are: Murder; Rape; Robbery; Assault

We will not be discussing Property Crimes such as *Burglary - breaking into and entering a private home with the intent to steal* and *Theft – stealing, most frequently motor vehicle theft, and larceny* because these crimes usually do not result in personal injury.

Murder: *First-degree murder* (colloquially referred to as murder one or in California 187 which is the state penal code) is premeditated homicide, or murder which is planned ahead of time whether it's years in advance or just a second before the act.

Second-degree murder or *voluntary manslaughter* is homicide done without forethought in the heat of the moment or in some states refereed to as adequate provocation.

Third-degree murder or *simple manslaughter* is homicide without the specific intent to kill. It may be the result of criminal negligence such as drunk driving or the result of an assault which culminated in death.

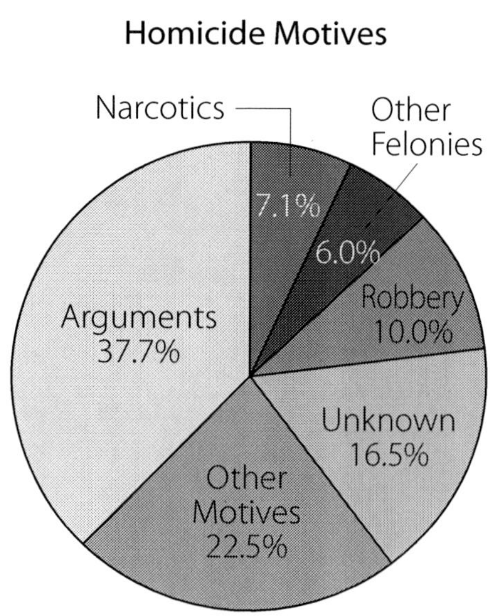

Top Ten Types of Weapons Used on Murder Victims

- Firearms
- Knives
- Blunt objects (hammers, clubs)
- Personal weapons (hands, fists, etc...)
- Poison
- Explosives
- Fire
- Narcotics
- Strangulation / Asphyxiation
- Drowning

Source: Crime in the United State, 2003, FBI, Uniform Crime Reports

COMMON SENSE STRATEGIES » *Murder*

- The highest percentage of murders are caused by heated arguments- never get trapped in such a situation – walk away for another day.
- Robberies can escalate to murder. Lives are priority, give up the money.
- Don't get involved with drugs.
- Stay away from unsavory characters and places.
- Never give someone a reason to benefit from your death.

Rape: A rape victim can be anyone – young children, grandmothers, students, mothers, wives, working women, old or young, rich and poor.

A rape crime can take place anywhere at any time – in high risk situations such as walking alone at night, jogging in an isolated park, frequenting bars late at night – or trapped in your own home, caught in the laundry room of your apartment building, walking down the street in broad daylight.

Stereotypes of rapists are wrong. They are not the sex-crazed, strange looking psychopathic strangers we usually imagine them to be. Too often, the rapist is a neighbor, acquaintance, friend or relative.

Facts:

- Sex is the weapon rapists use to exert control over the victim
- 75% of rapes are planned in advance
- Most rapists are heterosexual
- Most victims say that their primary concern was fear for their life
- Less than 3% of reported rapes turn out to be false reports
- Only 37% of all rapes are reported to the police

Resist or Submit:

Many variables factor into making the right choice for the right time. Each person must trust their instinct and capability to go forward with their decision. Be prepared for either option.

Resist: If you feel confident of attracting attention for help, escaping by talking your way out of the situation or you believe in your ability to use enough force against your attacker to get away, do it.

Submit: If you realize that your attacker will hurt you seriously or kill you if you put up a fight, then do what is necessary to survive.

Date Rape:

Date Rape is forced sex whether coerced by physical, emotional tactics or by drugs. Drugs are used to render the victim unconscious or impaired and less resistant to sexual assault. The drug will also limit the victim's memory. The most common drugs used by date rapists are alcohol, Rohypnol, GHB, and Ketamine. Since these substances, slipped into unsuspecting victims' beverages at dance parties or nightclubs, are tasteless and odorless, the victim has no way of knowing that the drink will leave her/him defenseless. The victim is ultimately taken to somewhere private and assaulted.

COMMON SENSE STRATEGIES » *Rape*

- If a situation or person makes you uneasy, leave immediately.
- Be prepared by knowing your options and capabilities.
- Always let someone know where you may be going.
- Make your home a safe haven. Think of your car as your portable home.
- Never let someone get you into their car – yell, scream, bite, kick.
- Avoid walking alone at night; avoid bushes, alleys, parking lots.
- Don't accept drinks from other people; don't share drinks.
- Keep your cell phone handy, have spare money for emergency.
- If you feel sick or dizzy while out socially, call a friend immediately.
- Practice the basics of "Rape Escape" or "CDT, A Life Training" program.

Robbery: The short definition of "robbery" is theft from a person by use of force or fear. Legally, the necessary elements are

- specific intent
- the act of theft from a person
- use of force or fear.

Facts:

- Most robberies involve only the victim and the robber and lasts less than a minute.

- Most robberies occur on the street (49%), then at late night retail businesses (24%).
- If a robbery is directed at an individual, this is referred to as a strong-arm robbery. If someone uses force to steal your car, it's called carjacking. If someone robs you by force in your home, it's called home invasion robbery. A shoplifter who fights with the store clerk is charged with the crime of robbery.
- The highest rate of robberies per capita occur in large metropolitan cities:

Large cities	514 per 100,000 population
National rate	165 per 100,000 population
Small cities	66 per 100,000 population
Rural areas	17 per 100,000 population

- Strong arm tactics like punching or threats are the most used weapon, followed by firearms and knives.
- Robberies occur usually between 8:00 P.M. and 2:00 A.M.
- Robbers prefer lone victims, few witnesses and good escape routes.

COMMON SENSE STRATEGIES *» Robbery*

- Everyone can't live in a rural area so we have to compensate by being more alert to the high risk of being robbed wherever we live.
- Robberies can escalate to murder. Lives are priority, give up the money.
- Don't go out alone at night especially between 8:00 p.m. and 2:00 a.m.
- Don't carry large amounts of cash or wear expensive jewelry in

casual environments.
- Insure your home and auto against theft.

Assault: Assault is the most common violent crime in the United States. The courts deal with assault as simple or aggravated.

Simple assault is the attempt or threat to inflict physical injury without a weapon.

Aggravated assault is "an unlawful attack by one person upon another for the purpose of inflicting severe or aggravated bodily injury, usually accompanied by the use of a weapon or by means likely to produce death or great bodily harm." (FBI)

Facts:
- The number of assaults has risen the past several years both nationally and in the cities, suburbs and rural areas.
- The Bureau of Census estimates that 93.8% of all assaults involve some type of weapon.
- Reports of assaults are evenly distributed between strangers (52%) and non-strangers (48%).
- The majority of assaults occur during the day (6:00 a.m. – 6:00 p.m.)
- 14% of all assaults occur in the home; 7% occur at a friend's, relative or neighbor; 3% in the street near victim's home.

Domestic Abuse: Domestic Abuse is battering, a pattern of behavior in a relationship that is used to keep control and power over another.

Types of Abuse: *Emotional, Psychological, Sexual, Physical*

Emotional:

- Attacks on you and people you care for, your family, friends
- Blaming you for everything that goes wrong
- Controlling everything you do and every penny you spend
- Telling you that you are sick and need counseling
- Treating you like a servant and degrading you
- Constant and repeated harassment
- Ridiculing your thoughts and beliefs

Psychological:

- Threatening to harm children and pets
- Threatening to harm you with a weapon
- Threatening to harm himself if you leave
- Hiding, stealing or destroying your possessions
- Sabotaging your car
- Throwing things, punching walls, slamming doors

Sexual:

- Forcing you to engage in sexual activity
- Against your will even though you say "no"
- When you are sleeping
- When you are afraid to deny him
- When you are unable to say "no"
- Criticizes you sexually – about your body, your dress, your history
- Flirts with other women and flaunts his affairs
- Forces you to do sexual acts you do not want to

Physical:

- Any forceful or violent physical behavior
- Using weapons of any kind
- Choking, slapping, punching, spanking, kicking,

shoving etc.

The Violence Wheel*4

The Violence Wheel is a representation of the Batterer's pattern of behavior.

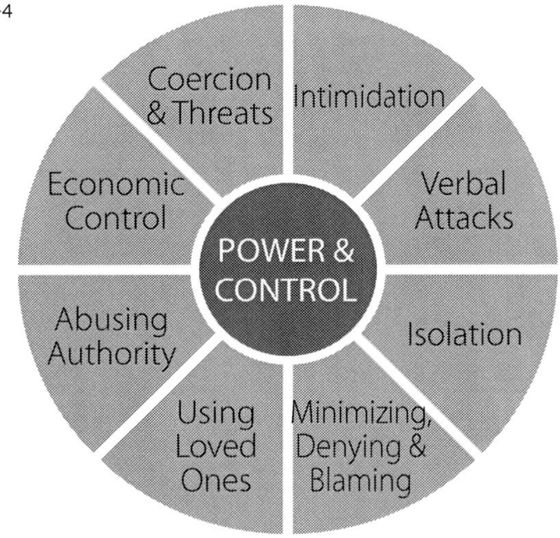

The Non-Violence Wheel*5

The Non-Violence Wheel presents a picture of a nourishing relationship.

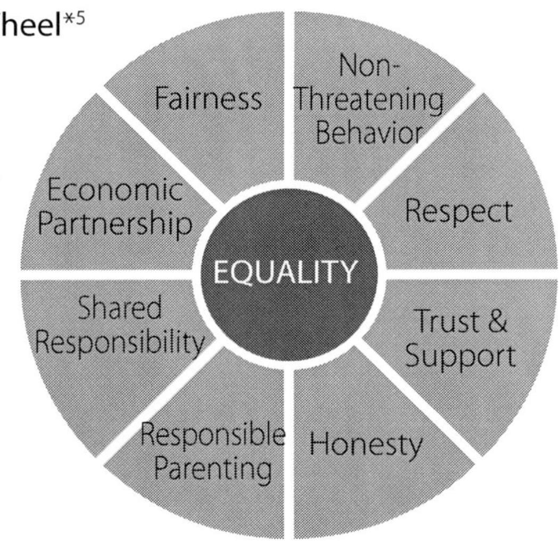

Facts:

- Many women are afraid to report domestic violence.
- Battering occurs by people of all faiths and all creeds.
- Battering will not stop by itself and becomes even more severe over time.

- Battering is dangerous and serious and can lead to severe physical harm.
- Men who are batterers are often loving, gentle and successful in other areas of their lives.
- Most men who are abusive are only violent at home. They are often friendly and well liked in public.
- Alcohol and drug use is often used as an excuse for violent behavior.
- Batterers are men from every race, religion, social, educational and financial background.
- Many women who are battered look normal and hide their injuries.
- Victims of abuse are often confused and overwhelmed. They tend to blame themselves because their self-esteem has been beaten into the ground.

Cycle of Domestic Violence*6

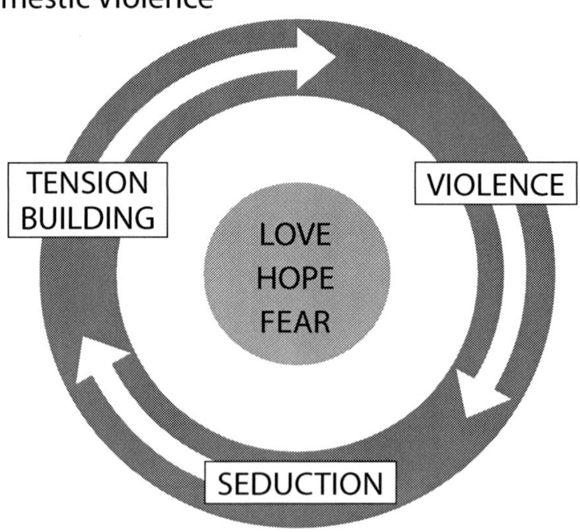

The Inner Circle: A Rollercoaster of Emotions

The victim is trapped in an unbroken cycle of three emotions:

- Love - tries to remember the good times, when he wasn't so bad.

- Hope - maybe he'll change if she forgives him.
- Fear - maybe he will act on his threats; maybe he won't change.

The Outer Circle: A Ring of Hidden Terror

The batterer moves from one phase to the other with only one goal – total control.

- Tension – threats and criticism
- Violence – physical, sexual and emotional attacks
- Seduction – lies, promises, gifts

COMMON SENSE STRATEGIES *» Domestic Abuse*

- Recognize the real problem and come out of denial.
- Make a decision to break the cycle of domestic violence.
- Ask for help from relatives, friends, people you trust.
- Talk to the Domestic Violence counselors in your area.
- Seek help from social services, emergency shelters.
- Plan your safety strategies.
- Form a support team to implement your plan including legal & financial advisors.

Fire

"Fire prevention is a whole lot easier than fire clean-up. You can unwet, but you can't unburn."

——— *Robert Dutton*

FIRE is a chemical chain reaction created by the combination of three elements illustrated by :

The Fire Triangle

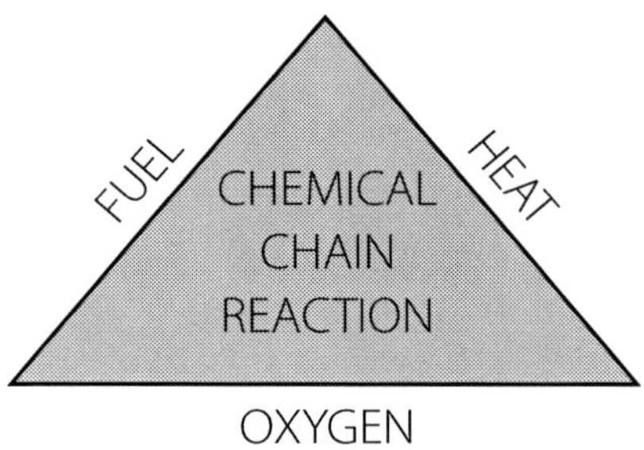

Fuel: The fuel may be a solid, liquid or gas or a combination of all three. The procedure for extinguishing a fire depends on the type and quantity of the fuel.

Oxygen: In an atmosphere of a minimum of 20 percent oxygen, fire will continue to burn as long as the fuel lasts.

Heat: Sufficient heat is needed to ignite the fuel material.

IMPORTANT NOTE:

Remove just one of the three elements and the FIRE will be extinguished.

Extingusihing Fires

Fire Type	Instrument	Removed
Class A: Solids paper, wood, cloth, plastic...	Water	Heat
	Foam	Heat & oxygen
	Dry chemical	Breaks chain reaction
Class B: Liquids gasoline, oil, kerosene...	Foam	Oxygen
	CO2	Oxygen
	Dry chemical	Breaks chain reaction
Class C: Electrical Equipment	CO2	Oxygen
	Dry chemical	Breaks chain reaction
Class D: Metals titanium, aluminum...	Special Agents	Oxygen

How to Use a Fire Extinguisher

Questions: Is it the right type of extinguisher?
Is it big enough to handle the size of the fire?
Is the area clear of dangerous materials?
Do you have an exit strategy?

Operate the portable fire extinguisher in an upright position.

Test that the extinguisher is working properly by doing the 4 steps:

P Pull safety clip

A Aim hose at the base

S Squeeze trigger

S Sweep back and forth at the source

Facts:

- Each year, more than 4,000 Americans die in fires. 25,000 are injured.

- The cause of death is usually smoke inhalation.
- Room temperature in a fire can be 100 degrees at floor level, but 600 degrees at eye level.
- A room can get so hot in five minutes that everything in it ignites at once, a phenomenon called "flashover."
- The most common cause of fires in rural areas and cold temperature areas is improper maintenance of chimneys.
- The most common cause of fires in all other areas is unattended cooking.
- Fires caused by burning candles are increasing.
- Faulty electrical wiring and worn appliance cords are another major cause of residential fires.
- Many people inhale the toxic gases from burning and die in their sleep.
- Having a working smoke alarm more than doubles your chance of surviving a fire.

COMMON SENSE STRATEGIES *» Fire*

- Clean chimney and any other heating appliance every season.
- Do not leave stove or oven on and unattended for any period of time.
- Blow out candles when celebration occasion is over.
- Practice fire safety drills with your children such as using ladders from the second story window or feeling your way out of the home with your eyes closed staying low to the ground.
- Make sure you have two exits from every room, windows should always be accessible.
- Check smoke detectors regularly.

- Obtain the fire extinguisher suitable for your situation.
- Follow your family escape plan.

NOTES:

Arson:

- The most common cause for arson is illegal profit. Other reasons include revenge, vandalism, concealment of another crime, and excitement.
- Arson is the second leading cause of residential fire deaths.
- Over 500,000 arson fires are reported each year

Arson is a serious crime. It not only destroys lives and property, it can affect entire neighborhoods by its indiscriminate attacks. It is our responsibility to report any suspicious activity.

Use an Insurance Professional

Find an insurance agent who works for a reputable company that you trust. Ask your agent to find the best policies appropriate for your situation. Then take advantage of discounts for multiple policies and combine auto, home, health and life policies together in your family portfolio.

Natural Disasters

NOTE: ✔ Check the category that applies to you.

	Earthquakes	Floods	Hurricanes	Landslides	Tornadoes	Tsunami	Volcanoes	Winter Storms
Alabama								
Alaska								
Arizona								
Arkansas								
California								
Colorado								
Connecticut								
Delaware								
Florida								
Georgia								
Hawaii								
Idaho								
Illinois								
Indiana								
Iowa								
Kansas								
Kentucky								
Louisiana								
Maine								
Maryland								
Massachusetts								
Michigan								
Minnesota								
Mississippi								
Missouri								

Reference Chart

	Earthquakes	Floods	Hurricanes	Landslides	Tornadoes	Tsunami	Volcanoes	Winter Storms
Montana								
Nebraska								
Nevada								
New Hampshire								
New Jersey								
New Mexico								
New York								
North Carolina								
North Dakota								
Ohio								
Oklahoma								
Oregon								
Pennsylvania								
Rhode Island								
South Carolina								
South Dakota								
Tennessee								
Texas								
Utah								
Vermont								
Virginia								
Washington								
West Virginia								
Wisconsin								
Wyoming								

Natural Disasters

> "In nature there are neither rewards nor punishments - there are consequences."
>
> ——— *Robert Ingersoll*

Man does not yet possess the technology to prevent the forces of nature from wrecking havoc and destruction on our cities, our homes and our environment. The best our scientists can do is give us as much advance warning as possible so that people can evacuate a danger zone. Our common goal is to save lives and, whenever possible, to reduce the level of property damage. In the case of a winter storm or a hurricane, meteorologists are able to track a storm's path days ahead of time. For a tsunami, the advance warning may be only a few hours. The devastation of a major earthquake can happen in minutes. Whatever the situation, if you follow the strategy of the POO Factor and utilize the Action Triangle, the chances for your family's safety will be greatly increased.

Refer to the *Natural Disasters Reference Chart*. Check off the disasters which happen in your neighborhood, your family and friends and also the places where you plan to visit.

PRIMARY OBJECTIVES:

- Prevent the event or avoid the danger.
- Preserve lives and reduce injuries.

SECONDARY OBJECTIVES:

- Prevent property damage
- Minimize losses and begin recovery as soon as possible

EARTHQUAKES

— What, Where, When —

An earthquake is a series of vibrations caused by a movement of part of the Earth's crust or plates caused by a sudden release of stresses.

There are hundreds of tremors each year. However, just five major earthquakes occurred in the United States in the previous century.

1906	San Francisco, California	(2000 lives lost)
1964	Alaska	(131 lives lost)
1971	San Fernando, California	(65 lives lost)
1989	Loma Prieta, California	(66 lives lost)
1994	Northridge, California	(61 Lives lost)

Areas with the greatest risk of a major earthquake include:

- The Western United States along the San Andreas Fault: **California, Washington State, Alaska**
- The New Madrid Fault Zone in **Missouri**
- Areas of **Massachusetts, North Carolina, South Carolina**

Earthquakes can happen at any time. There is no seasonal or annual cycle of occurrence. The vibrations may last for seconds or minutes. The aftershocks could happen even months after the main earthquake.

In order to understand the magnitude of an earthquake, look at the classifications based on the Richter Magnitude Scale*:

- M 1 to 3 Recorded but generally not felt
- M 3 to 4 Often felt, no damage
- M 5 Felt widely, slight damage near epicenter
- M 6 Damage to poorly constructed buildings within 10's km
- M 7 **Major** earthquake, serious damage up to 100 km
- M 8 **Great** earthquake, great destruction, loss of life
- M 9 **Rare Great** earthquake, major damage over large region

For the mathematicians who are interested, the Richter Scale has a logarithmic base which means every increment on the scale is multiplied by a factor of 10. For the rest of us, it may be easier to understand that the difference between an earthquake of the magnitude of 8.8 is not just twice as dangerous as one of 4.4 but 10,000 times worse.

IMMEDIATE DANGER:

- Falling debris, especially glass and heavy objects
- Unsafe Structures collapsing
- Fires, explosions, gas leaks

IMMINENT DANGER:

- Tsunami generated by earthquake
- Landslides and avalanches
- Dam failure triggering flash flood

CONTINUED RISKS:

- All Utilities disabled
- Roads, bridges and transportation systems damaged
- Emergency crews cut off

IMMEDIATE ACTION: When the tremor hits...

- **Drop, Cover and Hold.** If indoors, stay inside and find a safe spot such as under a desk or table.
- **Drop to the ground** if outside and find a spot away from buildings, trees, power lines and overpasses.
- **Buckle up** if in a vehicle and stop in a clear location.

CRUCIAL FOLLOW UP: When the shaking stops...

- Tune in to the Emergency Alert System (EAS) for emergency information and instruction.
- Activate your family emergency plan and phone tree.
- Based on the area you live, respond appropriately:
    - In high rise buildings, exit via the stairway.
    - In coastal areas and flood zones, move to higher ground.
    - On slopes, cliffs and mountainous areas, move to secure ground.

LONG TERM GOALS:

- Education and training to make your community more self-sufficient.
- Licensed professionals can install proper pipes to avoid gas and water leaks in order to prevent potential fires, explosions and flooding.
- Reinforce features in your home such as porches, decks, carports on the outside and cabinets, light fixtures and book shelves on the inside.

- What are three things you can do to increase your family's safety in the event of an earthquake?
  1.

  2.

  3.

- What are the safety areas and danger zones during an earthquake inside a building?

- Does your home insurance cover earthquakes?

FLOODS

— What, Where, When —

Whenever a body of water rises high enough to cover what is usually dry land, it is called a flood.

Floods are the most frequent of all the natural disasters and often are linked to other disasters such as earthquakes and hurricanes. For this reason, floods are ranked as the costliest natural disaster as high as 90% of the total economic losses suffered. From 1971 to 2000 the average number of lives loss per year was127. Damages cost an average of $4.1 billion each year in the same period. More than 4000 communities are at risk of flooding and it is estimated that in excess of 100 million people are potentially exposed to this danger.

Most parts of the United States are susceptible to some risk of flooding. However, the most vulnerable areas include:

- All along the **Mississippi River**
- The central valley of **California**
- Certain areas along the **Missouri** and **Arkansas** Rivers
- Floodplain areas in the **Gulf Coast**
- Widespread areas along the **South Atlantic**

Floods have many causes and may or may not be seasonal. For example, many people associate flooding with the annual spring snow and ice melt. Depending on the thaw, debris can cause a river or stream to overflow its banks.

Other critical factors must be considered such as:

- Rainfall intensity and duration
- Urbanization which increases runoff
- Dam and levee failure

IMMEDIATE DANGER:

- Flash flooding with little or no warning - six inches can knock a person off his feet.
- Debris in swift moving flood waters can cause serious injury.
- 24 inches of fast moving waters can sweep a car away.

IMMINENT DANGER:

- Bridges can be washed away without warning.
- Landslides and avalanches.
- If there is dam and levee failure, flooding can inundate towns and

residences downstream.

CONTINUED RISKS:

- All Utilities disabled.
- Roads, bridges and transportation systems damaged.
- Emergency crews cut off.

COMMON SENSE STRATEGIES » *Floods*

IMMEDIATE ACTION:

- Tune in to the Emergency Alert System (EAS) for emergency information and instruction.
- Activate your family emergency plan and phone tree.
- Evacuate as planned or instructed following safe route to shelter.

CRUCIAL FOLLOW UP:

- Do not drive through flood waters.
- Heed barricades! Avoid bridges, drains and irrigation ditches.
- Based on the area where you live, respond appropriately:
    - In coastal areas and flood zones, move to higher ground.
    - On slopes, cliffs and mountainous areas, move to secure ground.

LONG TERM GOALS:

- Education and training to make your community more self-sufficient.
- Double check with your insurance agent about coverage through the NFIP (National Flood Insurance Program).
- In order to protect your property, avoid building in a flood plain or make sure utilities and appliances are elevated.

- What are three things you can do to increase your family's safety in the event of a flood?

  1.

  2.

  3.

- How can you keep your important documents safe in the event of a flood?

- Does your home insurance cover floods?

HURRICANES

— What, Where, When —

A **hurricane** is defined as a storm of violent low pressure forming in the tropical Atlantic Ocean usually between the months of June through November. August and September are the peak periods. Hurricane force winds range between 75 miles per hour and 155 miles or more. Torrential rains, tornadoes, floods and storm surges usually are associated with major hurricanes.

Typhoons are similar storms which originate in the Western Pacific Ocean.

Coastal Storms may cause as much damage and pack similar force winds but are not classified as hurricanes simply because they do not originate in the tropics and are usually formed along the east coast from December through March.

Hurricane advance warnings have improved so much, the death toll has decreased dramatically. But property damage has increased as the population in high risk areas has increased. Approximately 100 million Americans live in areas which are prone to hurricane force winds. At least 14 million live in areas which have recorded winds greater than 125 mph. 6 million people or more live in storm surge locales.

Areas with the greatest risk of extreme hurricanes occurring include:

- The entire **Gulf Coast** and the **Southeastern Seaboard**.
- **Pennsylvania**, and the coasts of **New Jersey**, **New York** and **New England.**

Hurricane Classifications

Category	Windspeed (Miles per hour)	Storm Surge (feet)
1 Minimal	74 – 95	4 – 5
2 Moderate	96 – 110	6 – 8
3 Extensive	111 – 130	9 – 12
4 Extreme	131 – 155	13 – 18
5 Catastrophic	115 +	18 +

IMMEDIATE DANGER:

- Flying objects, branches, debris, broken glass
- Unstable structures collapsing

- Power lines down

IMMINENT DANGER:

- Be aware of the "eye" – the storm is not over
- Stay alert for tornadoes
- Possible flooding

CONTINUED RISKS:

- All utilities disabled: no water, no electricity
- Roads, bridges and transportation systems damaged
- Emergency crews cut off

COMMON SENSE STRATEGIES » *Hurricanes*

IMMEDIATE ACTION:

- Decide to **EVACUATE** or **STAY & SECURE**.
- Initiate your CPR: Communication, Plan, Resources.
- Board up all windows and doors.

CRUCIAL FOLLOW UP:

- Stay tuned in to the Emergency Alert System (EAS) for emergency information and instruction.
- Stay away from downed power lines or gas valves. Let the utility companies restore service.
- Do not enter areas until it is declared safe.

LONG TERM GOALS:

- Education and training to make your community more self-sufficient.
- Build a SAFE ROOM.

- Reinforce the structure of your house using hurricane clips and precut plywood for windows and doors.

▲ The POO FACTOR *» Hurricanes*

- What are three things you can do to increase your family's safety in the event of a hurricane?

  1.

  2.

  3.

- What are the safety areas and danger zones inside a building during a hurricane?

- Does your home insurance cover hurricane damage?

LANDSLIDES

— What, Where, When —

Landslides are a serious geologic threat in all 50 states. Every year approximately 25–50 people are killed and damages add up in the billions of dollars. Until we become smarter about the way we interact with the land, we will continue to suffer unnecessary and devastating losses.

Fast moving landslides are sometimes referred to as debris flows, mudslides, mudflows, lahars or debris avalanches. The most destructive debris flows occur after volcanic eruptions. During periods of intense rainfall or rapid snow melt, landslides that start on steep hillsides can exceed 35 miles per hour.

Gravity is the driving force which ultimately causes the landslide. However, the factors which set up the conditions for the landslide movement include the following:

- Over saturation – prolonged rainfall and flooding
- Alternate freezing and thawing
- Erosion because of construction or removal of trees
- Earthquake damage
- Volcanic eruption
- Burned out areas because of wildfires

Areas prone to landslides include:

- A history of previous landslides
- The bases of steep hillsides
- Areas where there are drainage channels
- Developed hillsides where leach-field septic systems are allowed

IMMEDIATE DANGER:

- Rocks, trees, and mud crashing on home
- Unsafe structures collapsing
- Fires, explosions, gas leaks because of damaged utilities

IMMINENT DANGER:

- Larger subsequent landslides generated
- Danger of more structures collapsing
- Possibility of more ground cave ins

CONTINUED RISKS:

- Shifting of foundations to buildings
- Increased erosion
- Utilities disabled

COMMON SENSE STRATEGIES *» Landslides*

IMMEDIATE ACTION:

- Evacuate away from the path of the landslide.
- Save yourself and your family, do not stop for possessions.
- If escape not possible, curl up into a tight ball and protect your head.

CRUCIAL FOLLOW UP:

- Tune in to the Emergency Alert System (EAS) for information.
- Check for trapped and injured victims.
- Report broken utilities.

LONG TERM GOALS:

- Check foundations and structures and repair damage.
- Seek expert advice on best ways to prevent or reduce landslide risk.
- Replant grounds in order to prevent further erosion.

▲ The POO FACTOR *» Landslides*

- What are three things you can do to increase your family's safety in the event of a landslide?

- Where do you seek shelter in a landslide?

- Does your home insurance cover landslides?

TORNADOES

— What, Where, When —

Tornadoes are circular windstorms. Wind speeds have been recorded in excess of 250 mph. Typically, tornadoes develop during severe thunderstorms and twisters have been documented as part of the danger present in hurricanes.

Official reports indicate that approximately 100 people are killed each year by tornadoes which number over 1000 annually.

Tornadoes have been sighted in every state but the areas with the greatest risk of tornadoes are:

- The Central United States from Kansas to Kentucky and from Texas to Minnesota. "Tornado Alley" includes Texas, Oklahoma and Kansas.

Tornado season lasts from March to August but can occur at any time throughout the year usually between 3:00 p.m. and 9: 00 p.m.

In order to understand the force of a tornado, look at the classifications based on the Fujita Wind-Damage Scale:

Wind-Damage Level	Wind Speed	Anticipated Damage
F 0	Up to 72 miles per hour	Light
F 1	73 – 112 mph	Moderate
F 2	113 – 157 mph	Considerable
F 3	158 – 206 mph	Severe
F 4	207 – 260 mph	Devastating
F 5	261 mph or greater	Incredible

Storm Advisory

- WATCH: means a tornado is possible in your area.
- WARNING: means a tornado has been sighted and may be headed for your area.

IMMEDIATE DANGER:

- Large hail
- Flying debris and broken glass
- Trees uprooted, cars and mobile homes overturned

IMMINENT DANGER:

- Damaged buildings might collapse
- Fallen power lines
- Ruptured gas pipes

CONTINUED RISKS:

- All Utilities disabled
- Roads, bridges and transportation systems damaged
- Undetected damage to roofs and structures of houses

COMMON SENSE STRATEGIES *» Tornadoes*

IMMEDIATE ACTION:

- Seek shelter in a safe room: basement away from all windows
- High rise building: a hallway in the center of the building
- Mobile home owners should seek shelter in a nearby safe building.

CRUCIAL FOLLOW UP:

- Tune in to the Emergency Alert System (EAS) for emergency information and instruction.
- Activate your family emergency plan and phone tree.
- Check on neighbors and those trapped in buildings.

LONG TERM GOALS:

- Education and training to make your community more self-sufficient.
- Licensed professionals can install proper pipes to avoid gas and water leaks in order to prevent potential fires, explosions and flooding.
- Work to improve more advanced warning systems in your area.

▲ The POO FACTOR *» Tornadoes*

- What are three things you can do to increase your family's safety in the event of a tornado?

- What are the safety areas and danger zones inside a building during a tornado?

- Does your home insurance cover tornado damage?

TSUNAMIS

— What, Where, When —

The word tsunami is a combination of the Japanese words "tsu" (which means harbor) and "nami" (which means wave). The phenomenon we call tsunami is a series of ocean waves usually generated by earthquakes or underwater landslides. Tsunami waves are unrelated to the tides – which are caused by the gravitational influences of the moon, sun and the planets.

A tsunami can occur at any time, and during any season. The first wave of a tsunami is usually not the largest in the series of waves. Depending on the different factors involved, one coastal area may experience no damaging waves and another community not very far away may be completely destroyed by the impact of the waves.

Tsunamis have killed over 350 people in Hawaii, Alaska and the West Coast and have caused millions of dollars of property damage. However, the devastation that occurred in the Sumatra Tsunami of December 26, when nearly 300,000 people from countries in the region and visitors from around the world were killed, underscores the importance of advance warnings and being prepared not only at home but when we are on traveling abroad.

Tsunami warnings are currently conducted by two agencies:

- The West Coast/Alaska Tsunami Warning Center (WC/ATWC) which is responsible for California, Oregon, Washington, British Columbia, and Alaska

- The Pacific Tsunami Warning Center (PTWC) which is responsible for issuing warnings to Hawaii, U.S. territories in the Pacific basin and to international authorities.

See map regarding areas with greatest potential for tsunamis:

United States Costal Areas Historically Subject to Tsumanis

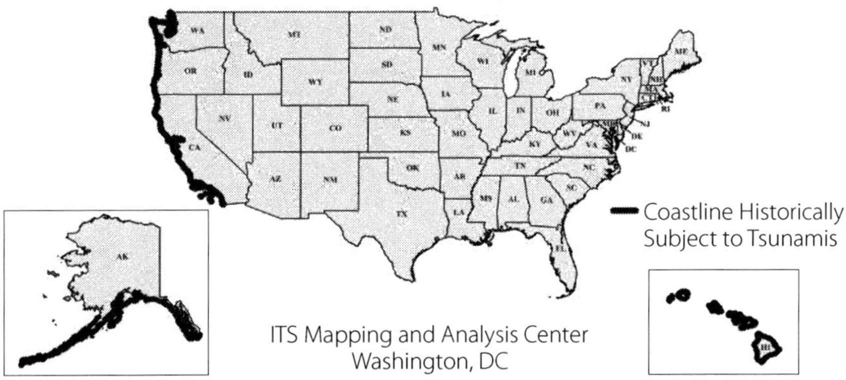

Source: National Oceanic and Atmospheric Administration, 2000

IMMEDIATE DANGER:

- Falling debris, heavy objects, broken glass
- Unsafe structures
- Fires, explosions, gas leaks from ruptured utilities

IMMINENT DANGER:

- Drowning
- The second wave may be larger than the first
- Contamination of drinking water

CONTINUED RISKS:

- Emergency crews cut off

- All Utilities disabled
- Loss of infrastructure

COMMON SENSE STRATEGIES » *Tsunamis*

IMMEDIATE ACTION: When the tremor hits...

- Drop, Cover and Hold. If indoors, stay inside and find a safe spot such as under a desk or table.
- Drop to the ground if outside and find a spot away from buildings, trees, power lines and overpasses.
- Buckle up if in a vehicle and stop in a clear location.

CRUCIAL FOLLOW UP: When the shaking stops...

- Tune in to the Emergency Alert System (EAS) for emergency information and instruction.
- Activate your family emergency plan and phone tree.
- Evacuate to higher ground.
- In high rise areas, evacuate vertically up to 6th floor.
- In coastal areas, move to higher ground as far inland as possible.

LONG TERM GOALS:

- Education and training to make your community more self-sufficient.
- Licensed professionals can install proper pipes to avoid gas ruptures and water contamination due to flooding.
- Rebuild foundation of buildings to prevent flooding

▲ The POO FACTOR » *Tsunamis*

- What are three things you can do to increase your family's safety in the event of a tsunami?

- Where do you seek safety when a tsunami threatens?

- Does your home insurance cover tsunamis?

VOLCANIC ERUPTIONS
— What, Where, When —

A volcano is nature's release valve through which molten rock escapes when the pressure from gases within the earth becomes too great. Lava flows consisting of molten rock can ooze out of the vent destroying everything in its path or erupt violently into the air creating fountains of red hot lava.

A unique hazard sometimes created by an eruption is a mudflow called a **Lahar**. Composed of a mixture of lava rock, debris and water from melting snow or intense rain, Lahars can flow down valleys and stream beds at speeds of 40 mph and travel more than 50 miles from it's original source.

The U.S. Geological Survey ranks Kilauea on the Big Island of Hawaii as the most dangerous in the nation followed by Mount St. Helens in Washington based on the past death toll estimates.

1790	Kilauea	(about 400lives lost)
1980	Mount St. Helens	(57 lives lost)

A recent review of 169 volcanoes in the United States and its territories recommend instituting a 24-hour, seven day Volcano Watch Office and overall increased monitoring. The four groups of volcanoes with the highest priority include:

Group 1	**Kilauea** and **Mauna Loa** in Hawaiian Islands
	Mount St. Helens in Washington State
	Mount Spurr in Alaska
Group 2	**The Cascade Range** – nine volcanoes
	Redoubt, Makushin, Akutan and Augustine in Alaska
Group 3	Nineteen volcanoes in **Alaska** and **Mariana Islands** that represent a high risk to aviation
Group 4	21 important but unmonitored volcanoes in **Hawaii Washington, Oregon, California, Alaska, Wyoming**

Japan, Indonesia and the United States have the most active volcanoes in the world. Beginning in 1980, at least five (5) volcanoes have erupted every year in the U.S.

IMMEDIATE DANGER:

- Molten Lava flow
- Explosive eruptions
- Toxic gases

IMMINENT DANGER:

- Earthquake
- Tsunami generated by earthquake
- Landslides and avalanches

CONTINUED RISKS:

- Volcanic ash can collapse roofs
- Contaminate water supply
- Cause severe respiratory problems

COMMON SENSE STRATEGIES *» Volcanic Eruptions*

IMMEDIATE ACTION:

- Evacuate eruption area immediately.
- Hold a damp cloth over your face to help you breathe.
- Stay away from volcanic ashfall areas.

CRUCIAL FOLLOW UP:

- Tune in to the Emergency Alert System (EAS) for emergency information and instruction.
- Activate your family emergency plan and phone tree.
- Based on the area you live, respond appropriately in case of tsunami.

LONG TERM GOALS:

- Clear roofs of ashfall
- Reinforce features in your home such as porches, decks, carports
- If anyone in your family has a respiratory ailment, seek medical advice and stay indoors until health officials give the all clear

▲ The POO FACTOR *» Volcanic Eruptions*

- What are three things you can do to increase your family's safety in the event of a volcanic eruption?

- After an eruption, debris and ash will be carried by the wind and gravity. Which areas should you avoid so as not to be exposed?

- Does your home insurance cover roof damage caused by volcanic ash?

WINTER STORMS

— What, Where, When —

Winter season brings with it beautiful images of white, soft-fallen snow, sleigh bells ringing on Clydesdale horses pulling sleds full of presents to be placed under the Christmas trees decorated in cozy living rooms.

Winter also brings with it:

- Strong winds
- Extreme cold
- Ice storms
- Snow storms
- Sleet
- Freezing rain

It doesn't matter where you live. Some time during the season, winter's impact will affect you in one way or another. From the Mid-Atlantic Coast to New England, along the Gulf Coast and Southeast, in the Midwest and Plains, from the Rockies to the West Coast, in Alaska and in Hawaii, contingency plans should be made on how to deal with winter weather.

The Dangers of Winter Storms:

Snow Fall:

- *Snow showers:* snow falling for short period in various intensity with little or no accumulation
- *Snow squalls:* brief but intense showers accompanied by gusty winds producing significant accumulation
- *Blowing snow:* wind driven snow causing reduced visibility
- *Blizzards:* winds of 35 mph or more with blowing snow reducing visibility to less than one-quarter mile for at least 3 hours

ICE:

- *Freezing rain:* rain that falls when temperatures are below freezing causing it to freeze on those surfaces and cause a significant danger.
- *Sleet:* raindrops that freeze into ice pellets before reaching the ground are called sleet. Sleet usually does not stick to the surface but accumulates like snow.
- *Ice storm:* ice storms occur when freezing rain falls ands freezes immediately. This is often the cause for power and communication disruption.

COLD:

- *Wind chill:* wind chill is not the actual temperature, but the estimate on how wind and cool impacts exposed skin.
- *Frostbite:* frostbite is damage to body tissue caused by extreme cold which results in a loss of feeling. Victims require medical attention as soon as possible.
- *Hypothermia:* hypothermia can occur when the body temperature falls below 95 degree Fahrenheit. Hypothermia is a very dangerous condition.

Winter Storms are called **Deceptive Killers** because most deaths are only indirectly caused by the storm:

- People die from the extreme cold when they are stranded in snow storms.
- People die of hypothermia because their homes have no power for heat.
- People die on the road in traffic accidents caused by icy roads.
- People die because of accidents caused by the strong winds.
- People die in collapsed buildings caused by heavy snow.
- People drown in the flooding caused by ice jams on the rivers.

Storm Advisories: The National Weather Service (NWS) Storm Prediction Center issues watches and warnings of hazardous weather.

Winter Storm Watch:
Severe winter conditions, such as heavy snow or ice, are possible within the next day or two. Prepare now.

Winter Storm Warning:
Severe winter conditions have begun or are about to begin in your area. Stay indoors!

Blizzard Warning:
Snow and strong winds will combine to produce a blinding snow (near zero visibility), deep drifts, and life threatening wind chill. Seek refuge immediately!

IMMEDIATE DANGER:

- Motor vehicle accidents
- Hypothermia and asphyxiation

- Exhaustion and heart attacks

IMMINENT DANGER:

- Disruption of medical services
- Potential house fires
- Stranded travelers outdoors

CONTINUED RISKS:

- Power and communication lines disrupted
- Emergency repair crews cut off
- Airports closed

COMMON SENSE STRATEGIES *» Winter Storms*

IMMEDIATE ACTION:

- Listen to NOAA weather radio for changing weather conditions.
- Activate your family plan; restock your emergency supply kit.
- Avoid unnecessary travel.

CRUCIAL FOLLOW UP:

- Stay indoors during the storm.
- Keep warm; eat and drink regularly to maintain energy.
- If you must go outside, understand the dangers of hypothermia and overexertion.

LONG TERM GOALS:

- More training in emergency skills like CPR and First Aid.
- Obtain equipment like AED or a portable generator to make your family more self-sufficient.
- Reinforce features in your home such as the roof and decks to

prevent collapse from heavy snow fall.

▲ The POO FACTOR *» Winter Storms*

- What are three things you can do to increase your family's safety during severe winter storms?

- Where is the best place to seek shelter in a winter storm?

- Does your home insurance cover collapsed roofs due to heavy snow?

Use a Real Estate Professional

When purchasing a home or property, rely on your Real Estate Agent to get documentation about the history of the area you are considering. A complete disclosure should include a record of landslides, floods, earthquakes, wildfires, tsunamis and anything else which may influence your decision.

Terrorism

> "We are a different Nation today: sadder and stronger, less innocent and more courageous, more appreciative of life."
>
> ——— *President George W. Bush*

Terrorists don't put a value on human life, no one's - including their own. That's the difference between us. We care about our families. We care about our neighbors. We care about our communities. We care about our country. We care. And that's why, in the end, we will win the fight against terrorism.

Excerpt from Author's Note:
... almost everybody we talked to expressed the feeling that for them the world had changed on 9/11. Then they stressed that they were more aware now, more appreciative of their family and friends. And they noticed that more people seemed to be looking out for one another. Here's an example from a conversation we had with a room attendant at the Waldorf –Astoria in New York city. "You know, I was sick one day and couldn't report to work. Well, you wouldn't believe it," she said, grinning mischievously from ear to ear, "but my boss called and talked to my husband." She paused to make sure we understood how unusual this was. "You know why he called?," she asked. Not waiting for an answer she proudly continued, "he just called to see if I was o.k.? Imagine that!"

TERRORISM as defined by the U.S. Department of Justice is...

"the unlawful use of force or violence committed by a group or individual against persons or property to intimidate or coerce a government, the civilian population, or any segment thereof, in furtherance of political or social objectives."

TERRORISTS (foreign or domestic; individual or group) use the fear of uncertainly as one of their main strategies.

Major Terrorist Incidents In The United States

1993	Bombing of the World Trade Center
1995	Bombing of the Murrah Federal Building in Oklahoma City
1996	Bombing at the Atlanta Olympic Ganes
1996	Bombings at family planning clinics in Atlanta area
1997	Bombings at gay bars in Atlanta area
2001	Destruction of World Trade Center in New York Attack on the Pentagon
2001	Sending Anthrax through the U.S. Mail

Terrorists Targets: Each terrorist group will identify targets which match their specific goal. Some of the intended goals include:

- Panic and confusion
- Mass destruction
- Attack on America's top leaders and institutions
- Disruption of the economy
- Attack on crucial resources and services

Explosives: Most commonly used weapons by terrorists. Usually referred to as IEDs (improvised explosive devices), the materials required for such explosives can be purchased commercially (for example, ammonium nitrate which is also used as fertilizer). The raw materials can also be obtained from construction blasting supply companies or even manufactured using household ingredients. This type of explosive is most likely used in the back packs of suicide bombers.

Incendiary Devices: A device with three basic components: a fuse, a container and the material or fuel to start a fire.

Chemical Weapons: Terrorists can turn common, household products into lethal weapons. The ingredients used to manufacture chemical weapons are not difficult to produce or purchase. Chemical weapons include:

- Blister agents: causes burns and tissue damage
- Blood agents: absorbed into the blood stream causing the victim to lose consciousness.
- Choking agents: attacks the lungs; death occurs from lack of oxygen
- Nerve agents: affects the nervous system. Sarin is an example of a nerve agent which acts within seconds of exposure and cause convulsions and inability to breathe
- Riot-control agents: deigned to incapacitate like tear gas or pepper spray

Biological Weapons: The ability of science to tap into nature's secrets can be used for good or evil. Terrorists could use biological agents as a weapon to destroy large segments of the population and contaminate the food supply. Many biological agents are contagious and symptoms would take days or weeks to be noticed.

Radiological Devices: Hospitals, science laboratories, medical facili ties, clinics and many commercial companies use radiological materials. It is possible that terrorists could make an RDD (radiation dispersal device) using small amounts of radiation obtained from an x-ray machine.

Nuclear Weapons: The difference between a conventional weapon and a nuclear attack is the long term effects caused by contamination over such a wide ranging area. The number of people killed and injured in the initial attack would be just the beginning in the final death toll. At this time, experts believe the terrorists do not have the capability to use nuclear weapons.

COMMON SENSE STRATEGIES *» Terrorism*

- Listen to Emergency Alert System for news bulletins and instructions.
- Develop your family communication plan.
- Update your disaster supply kit.
- Practice shelter-in-place procedures.
- Designate the safe room in the house, at work and at school.
- Designate a safe meeting place outside of the home.
- Identify potential targets in your community.
- Imagine the worst case scenario and prepare an action plan.
- Be aware of environmental changes which could indicate a biological or chemical attack.

▲ POO FACTOR INDEX

DANGER	WARNING	FREQUENCY	POTENTIAL DAMAGE	PRIORITY
Hurricane	3 Minimal 2 12 hours 1 24 hours	3 High 2 Medium 1 Low	3 High 2 Medium 1 Low	3 High 2 Medium 1 Low
Tornado	3 Minimal 2 12 hours 1 24 hours	3 High 2 Medium 1 Low	3 High 2 Medium 1 Low	3 High 2 Medium 1 Low
Fire	3 Minimal 2 12 hours 1 24 hours	3 High 2 Medium 1 Low	3 High 2 Medium 1 Low	3 High 2 Medium 1 Low
Terrorist Attack	3 Minimal 2 12 hours 1 24 hours	3 High 2 Medium 1 Low	3 High 2 Medium 1 Low	3 High 2 Medium 1 Low
Kidnapping	3 Minimal 2 12 hours 1 24 hours	3 High 2 Medium 1 Low	3 High 2 Medium 1 Low	3 High 2 Medium 1 Low
Car Accident	3 Minimal 2 12 hours 1 24 hours	3 High 2 Medium 1 Low	3 High 2 Medium 1 Low	3 High 2 Medium 1 Low
Heart Attack	3 Minimal 2 12 hours 1 24 hours	3 High 2 Medium 1 Low	3 High 2 Medium 1 Low	3 High 2 Medium 1 Low

Did you know? ❶

The largest landslide in the world in this century occurred at Mount St. Helens, Washington, in 1980. An earthquake caused roughly 1.7 cubic miles of rocks and mud to break free and slide down the side of the volcano.

Did you know? ❷

The largest earthquake this century was a magnitude 9.5 event that struck Chile on May 22, 1960. More than 2000 people were killed in Chile, Hawaii, Japan, and the Philippines from the earthquake and tsunami generated.

Did you know? ❸

The largest eruption in the world this century occurred June 6-9, 1912 at Novarupta on the Alaska Peninsula. An estimated 9 cubic miles of magma was explosively erupted over 60 hours.

PART 3

Quick Reference/ Tips at a Glance

"No community is truly prepared for a disaster until every individual, family and household takes personal responsibility for preparedness"

——— *Marsha J. Evans, President American Red Cross*

Directory of Quick Tips

Category	QT	Tip	Page
MEDICAL	QT 1	Requirements for the Emergency Room	163
	QT 2	Choking-Prevention Tips	164
	QT 3	First Aid for Poisoning	165
	QT 4	How to Fight the Flu	166
	QT 5	Reminders for Diabetics	167
ACCIDENT	QT 6	How to Spot a Drunk Driver	168
	QT 7	Preventing Falls from High Rises	169
	QT 8	If You're Around Water	170
	QT 9	Safety Tips for Swimming Pool Owners	171
	QT 10	If Your Airplane Crashes	172
MISSING PERSON	QT 11	If Your Child Is Missing	173
	QT 12	Ten Suggestions for Parents	174
	QT 13	Stranded in Snow	176
VICTIMS	QT 14	Ten Tips to Avoid Shark Attack	177
	QT 15	Safeguard Against Assault	178
	QT 16	Protect Yourself Against Roofies, Extasy and Special K	179
	QT 17	Ten Warning Signs of Potential Drug Problems	180
FIRE	QT 18	If There's a Fire	181
	QT 19	Wildfire	182
NATURAL DISASTER	QT 20	Which Plan Should You Take	183
	QT 21	Evacuate	184
	QT 22	Stay & Secure	185
	QT 23	Earthquake	186
	QT 24	Hurricane Watch	187
	QT 25	Hurricane Warning	188
	QT 26	Stay or Go	189
	QT 27	Don't Forget Your Pet Plan	190
TERRORISM	QT 28	Homeland Security Advisory System	191
	QT 29	Riding On Public Transportation	193

Requirements for Emergency Room

- Family member, caregiver, friend who can help
- Description of emergency symptoms
- Current Medications: prescribed, over the counter and supplements
- Bring list or samples of all plus record of dosages.
- List of allergies: foods, drugs, etc.
- Record of medical conditions: diabetes, asthma, cancer, arthritis, lupus, etc.
- Record of prior surgeries
- Names and phone numbers of family physicians and specialists
- Insurance Information: Company, Policy Number, Agent
- Copy of living will
- Name and phone number of your representative with Power of Attorney

NOTE: EVEN IF YOU HAVE NO HEALTH INSURANCE, THE LAW REQUIRES THAT YOU RECEIVE PROPER TREATMENT.

Choking – Prevention Tips

- Recognize choking: an infant cannot cry or make normal sounds, infant may look blue or frightened, watch for silent cough.
- Don't leave small objects, such as buttons, coins and beads within an infant's reach.
- Have children sit in a high chair or at a table while they eat.
- Do not let children eat too fast.
- Make sure toys are too large to be swallowed.
- Do not give infants and small children foods such as nuts, grapes, popcorn or raw vegetables. Avoid hot dogs or cut them into round "coins" and then cut the coin slices in fourths.
- Make sure toys have no small parts that can be pulled off.
- Supervise children when they eat.
- Sign up for CPR and First Aid classes.

Source: American Red Cross

First Aid for Poisoning

Has the person collapsed or stopped breathing?

- Call 911 right away.

Poison in the Eyes?

- Rinse eyes with running water for 15 to 20 minutes.
- Call Poison Center and/or your doctor.

Poison on the Skin?

- Take off any clothing that the poison touched.
- Rinse skin with running water for 15 to 20 minutes.
- Call Poison Center and/or your doctor.

Inhaled Poison?

- Get to fresh air right away.
- Call Poison Center and/or your doctor.

Swallowed the wrong medicine or too much medicine?

- Call Poison Center and/or your doctor.

Swallowed something that's not food or medicine?

- Drink a small amount of milk or water.
- Call Poison Center and/or your doctor.

How To Fight The Flu

- Get an annual flu shot in the fall especially for those aged 50 and older or those with a chronic heart problem.
- Consult with your doctor about preventing flu with antiviral medications.
- Always call your doctor at the first sign of flu symptoms.
- Learn to recognize flu symptoms in different age groups:

  Children younger than 5: fever, vomiting, diarrhea, sore throat

  Children older than 5: fever, cough, chills, sore throat, headache, fatigue

  50 and older: fever, confusion, fatigue, nasal inflammation
- Keep your distance from people with the flu.
- Wash your hands frequently.
- Drink plenty of liquids.
- Maintain a good diet, get plenty of rest and control stress.
- Flu is contagious for 3-4 days after symptoms appear. Keep your children home from school and/or stay home from work if you have the flu.
- Remember that viruses are usually spread through the mouth, nose and eyes.

Reminders For Diabetics

Especially on Trips Outside of the Country

- Wear a medical ID bracelet concerning your diabetes.
- Pack extra diabetes medication and include enough in your carry on bag for easy access.
- Make sure you bring in your emergency kit: diabetes pills, insulin and syringes, blood and urine testing supplies, extra batteries for your glucose meter.
- Bring Medical insurance card and emergency SOS evacuation.
- Have your doctor's prescription for insulin or diabetes pills.
- Carry information about what to do in case of emergency and inform anyone traveling with you about your condition in case of emergency.
- Know how to adjust your meal plan or medication.
- Bring snacks for quick sources of sugar.
- Follow your normal schedule for checking your blood sugar.
- If you eat exotic foods, find out if it's starch, vegetable, meat or fruit and adjust your meal plan.
- Have your doctor's contact information and best way to reach him.

How To Spot a Drunk Driver

- Straddling lanes or driving on the center line.
- Drifting or moving in a straight line at a slight angle to the roadway.
- Driving with headlights off at night.
- Erratic braking or stopping without cause.
- Driving below the speed limit.
- Slow response to traffic signals.
- Sudden stops, delayed starts.
- Nearly striking an object, curb, sign...
- Weaving or zigzagging across the road.
- Driving on the wrong side of the road or completely off the road.
- Tailgating.
- Take down the vehicle description, the license plate number, the direction traveling and the street name. Report to police. Do not attempt to stop the vehicle.

Source: Adapted from MADD

Preventing Falls from High-rises

- Constant supervision is the key to preventing falls among children.
- Lock doors to balconies.
- Keep windows closed when children are around. If you need to open a window, open only those that a child cannot reach.
- Consider various types of alarms that alert you when sliding doors and windows are opened.
- Install window guards on all windows above the first floor excluding emergency fire exits. Guards should have a quick release mechanism that adults can open easily in an emergency.
- Do not rely on screens to prevent falls.
- Keep children away from all open windows.
- Install approved window locks and make sure everyone in the family knows how to open them in case of an emergency.
- Do not nail windows shut.
- Keep furniture and other items that children can climb on away from balconies, guardrails and open windows.

Source: Emergency Medical Services

If You're Around Water

Always check for a **LIFEGUARD** whether you are at the beach, the lake, the pool or any other swimming area.

- Take swimming lessons.
- Always swim with a grown–up.
- Learn and observe the water safety rules.
- Swim only when you are rested.
- Don't eat candy or chew gum when you are swimming. You could choke.
- Never jump or dive into the water unless it's approved for diving.
- Don't swim at night.
- If you hear thunder or see lightning, get out of the water.
- Don't listen to dares.
- Gates surrounding swimming pools should be locked to protect younger children.
- Never Boat alone.
- Always wear an approved PFD (Personal Flotation Device).
- Sitting or standing on the edge can cause a boat to tip over.

Safety Tips for Swimming Pool Owners

- Teach your children how to swim and how to exit pool.
- Never leave your children alone in or near the pool, even for a moment.
- Put up a fence to separate your house from the pool. Most young children who drown in pools wander out of the house and fall into the pool. Install a fence at least 4 feet high around all 4 sides of the pool. Use gates that self-close and self-latch, with latches higher than your children's reach.
- A power safety cover that meets the standards of the American Society for Testing and Materials (ASTM) adds to the protection of your children.
- Keep rescue equipment (such as a shepherd's hook or life preserver) and a telephone by the pool.
- Do not let your child use air-filled "swimming aids" because they are not a substitute for approved life vests and can be dangerous.
- Anyone watching young children around a pool should learn CPR and be able to rescue a child if needed. Stay within an arm's length of your child.
- Remove all toys from the pool after use so children aren't tempted to reach for them.
- After the children are done swimming, secure the pool so they can't get back into it.

Source: American Academy of Pediatrics

If Your Airplane Crashes

Tips For Travelers

1. Know where the exits are in front and behind

    Count how many rows to closest exit.

2. Use the crash position

    Brace for impact with head against seat in front of you.

3. Leave your luggage and things behind

    A Gucci Bag is not worth dying for nor is a laptop.

4. Keep low and breathe slowly

    Breathing aircraft-fire smoke will hurt your lungs.

5. Keep your shoes on

    Exit quickly. The slides are tougher now. Help at the bottom.

Passengers can do certain things to increase the odds of walking away from a crash unharmed. Survival isn't necessarily a matter of fate.

"The flying public thinks if you're in an accident, you're going to die. So you don't need to know what to do, and don't pay attention to the briefing or read the safety card. Nothing could be farther from the truth."
David Palmerton, Federal Aviation Administration expert on protection and survival.

If Your Child Is Missing

First, check everywhere in the house – wherever a child may crawl or hide. Check with your neighbors and your child's friends.

Next, call **911** to file a report with the police and ask about the potential of initiating the **AMBER ALERT**.

Request that the police enter your child's name and identifying information immediately into the **National Crime Information Computer (NCIC) Missing Person File**.

Contact your local Sheriff and **Search & Rescue (K9)** Department

Contact FBI (Missing Children Task Force)

Call the **National Center for Missing and Exploited Children 1-800-843-5678 (1-800-The-Lost)**

Call **Team Hope 1-866-305-4673 Parent Support Network**

Contact local media (television, radio, newspapers)

When you call law enforcement, provide the information in your **CHILD ID PROFILE** of **THE RED BOOK**. Provide your Child's Name, Date of Birth, Height, Weight, any other identifying characteristics and disabilities. Tell them what he/she was wearing and when you noticed your child was missing. The most powerful tool to locate a missing child is a current photo.

If your child disappears in a store, notify manager immediately. Many stores have an action plan called **CODE ADAM** – store employees immediately lock down and mobilize to find the missing child.

REMEMBER

There is no waiting period required to report a missing child.

10 Suggestions For Parents

1. Know your child's friends.

   Encourage your child to bring friends to your home. These friends will often be involved with you and your child for the rest of your lives.

2. Visit your child's school.

   Let your child know that you are interested and involved in your child's education.

3. Set aside time each evening when you and your child can discuss the day's events.

   This will help keep you in tune with your child's activities.

4. Encourage your child to discuss with you anything of concern or interest.

   There is nothing too small or insignificant for a child to discuss with you.

5. Make sure that your child knows daily that he/she is loved.

   Consider giving your child a hug and telling him you love him.

6. Set consistent standards of behavior for your child.

   Make sure that your child has a clear understanding of your expectations.

7. **Reward and discipline your child.**

   Let your child know that there are consequences for acceptable and unacceptable behavior.

8. **Set a good example.**

   Display the type of behavior you want to encourage.

9. **Discuss important values and attitudes with your child.**

   Good ethical grounding will help sustain a parent and a child through the good and bad times.

10. **Share a dream.**

    Setting realistic goals will help your child control his destiny.

——— Judge Michael Town

Stranded in the Snow

- Use a cellular telephone to call for help.
- Get out your winter storm survival kit
- Use flares or reflective triangles to attract attention and to avoid being hit by another car.
- Never walk away from your vehicle unless you know exactly where you are, how far it is to help and if you are positive you can improve your situation.
- Leave a note with date, time and destination in the car.
- Keep at least one window open slightly.
- Keep snow and ice from blocking the exhaust pipe.
- Run the engine and heater for 10 minutes every hour, depending on how much gas you have in your tank.
- Don't panic. Stay calm and reason each step through carefully.

Source: National Safety Council

10 Tips to Avoid Shark Attack

1. Swim, surf or dive with other people, and don't swim too far away from help.
2. Stay out of the water at dawn, dusk and night, when some species of shark move inshore to feed.
3. Do not enter the water if you have open wounds or are bleeding in any way. Sharks can detect even small amounts of blood and body fluids.
4. Avoid murky waters, harbor entrances, and areas near stream mouths (especially after heavy rains), channels or steep drop-offs. These types of waters are known to be frequented by sharks.
5. Do not wear high-contrast clothing or shiny jewelry. Sharks can see contrast well.
6. Refrain from excessive splashing. Keep pets, which swim erratically, out of the water. Sharks are known to be attracted to such activity.
7. Do not enter the water if sharks are known to be present, and leave the water quickly and calmly if one is sighted. Do not provoke or harass a shark, even a small one.
8. If fish or turtles start to behave erratically, leave the water. Be alert to the presence of dolphins, as they are prey for some large sharks.
9. Do not swim near people fishing or spearfishing. Stay away from dead animals in the water.
10. Swim or surf at beaches with lifeguards and follow their advice.

Safeguard Against Assault

- Stand tall and walk with confidence. Watch where you are going and what is going on around you.

- Walk along well-lit and busy streets. Walk with friends. Avoid shortcuts, dark alleys, deserted streets and wooded areas.

- Know your neighborhood. Identify police stations, libraries, schools – as well as the hours of operation of local stores and restaurants.

- Don't carry more money than you need for the day, but do carry emergency change for a telephone call.

- When you are out late at night, have a friend accompany you – don't go alone. Also, let someone know where you will be going and when you will return.

- Never hitchhike.

- When driving, always park in well-lit places and lock your doors.

- Before entering your vehicle, check for offenders hiding in the back seat or on the floor.

- If harassed or assaulted, scream and attempt to run to safety.

Source: National Crime Prevention Council

Protect Yourself Against Roofies, Extasy and Special K

- Don't go out by yourself.
- Never accept a drink from someone you don't know and trust.
- Monitor the behavior of friends who seem more intoxicated than they should be.
- If you hear someone just "joking" about date rape drugs, leave that party or individual.
- Never leave your drink unattended.
- Don't share or exchange drinks.
- Don't take drinks from a large open container or punchbowl.
- If you feel you are becoming intoxicated after only one or two drinks, get help.
- Beware of someone who wants to take you out for fresh air especially someone you just met.
- Consider using Drink Safe Technology products.

10 Warning Signs of Potential Drug Problems

1. Drop in academic performance
2. Lack of interest in personal appearance
3. Withdrawal, isolation, depression, fatigue
4. Aggressive, rebellious behavior
5. Hostility and lack of cooperation
6. Deteriorating relationships with family
7. Change in friends
8. Loss of interest in sports and/or hobbies
9. Change in eating/sleeping habits
10. Evidence of drugs or drug paraphernalia
    (e.g. needles, pipes, lighters, bruises, needle marks, red eyes)

Source: D.A.R.E. "Keeping Kids Drug Free"

If There's a Fire

If there's smoke and fire...

- STAY LOW & GO...

  Crawl on your hands and knees

  NOT your belly

If your clothes catch on fire...

- STOP...

  Don't Run! Running makes the fire worse.

- DROP...

  Drop to the floor on your hands and knees and then lie flat on your stomach.

- ROLL...

  Cover your nose, eyes and mouth with your hands and roll back and forth. This will make the fire go out.

- COOL...

  Cool the burn with clean, clear, fresh water after the fire is out.

- CALL...

  Call 911 for help from your neighbor's house.

Safety Zone to Prevent Brush Fires

- Mow the grass regularly.
- Rake and clear combustible vegetation such as leaves and twigs.
- Remove leaves and rubbish from under structures.
- Remove dead branches that extend over the roof.
- Remove vines from the walls of buildings.
- Clear an area around propane tanks and the barbecue.
- Dispose of newspapers and rubbish regularly.
- Store gasoline, oily rags and other flammable materials in a safe location.
- Have a garden hose long enough to reach all structures on the property.

Wildfire Emergency

- Make sure everyone knows how to call 911.
- Show family members how to shut off water, gas and electricity.
- Plan several escape routes away from home by car and foot.
- Pick meeting sites in case family members are separated.
- Make plans to take care of children in the neighborhood.
- Make list of items to bring if advised to evacuate.

WHICH CHOICE SHOULD YOU MAKE?

EVACUATE

- Fire
- Tsunami
- Flood
- Earthquake
- Landslide
- Volcanic Eruption
- If you are advised to leave by appropriate officials.

PLAN

PREPARE

PRACTICE

STAY & SECURE

- Hurricane
- Tornado
- Winter Storm
- Thunder Storm
- If you have a basement.
- If you have a safe room.
- If there is no shelter available.

EVACUATE

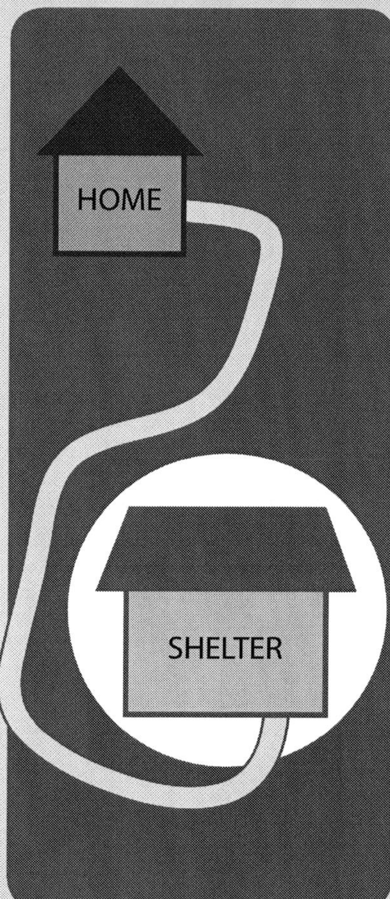

STEP 1

Communication:

- Use your NOAA weather radio
- Notify family and friends of your plans
- Keep cell phone charged

STEP 2

- Assemble your evacuation kit
- Bring important paperwork

STEP 3

- Initiate Pet Plan

STEP 4

- Bring extra cash
- Bring pillows, blankets
- Bring toys, games, books

STEP 5

- Secure Home

STEP 6

- Map out safe route to destination
- Fill your car with gas

STEP 7

- Travel out of danger zone or to Safe Shelter

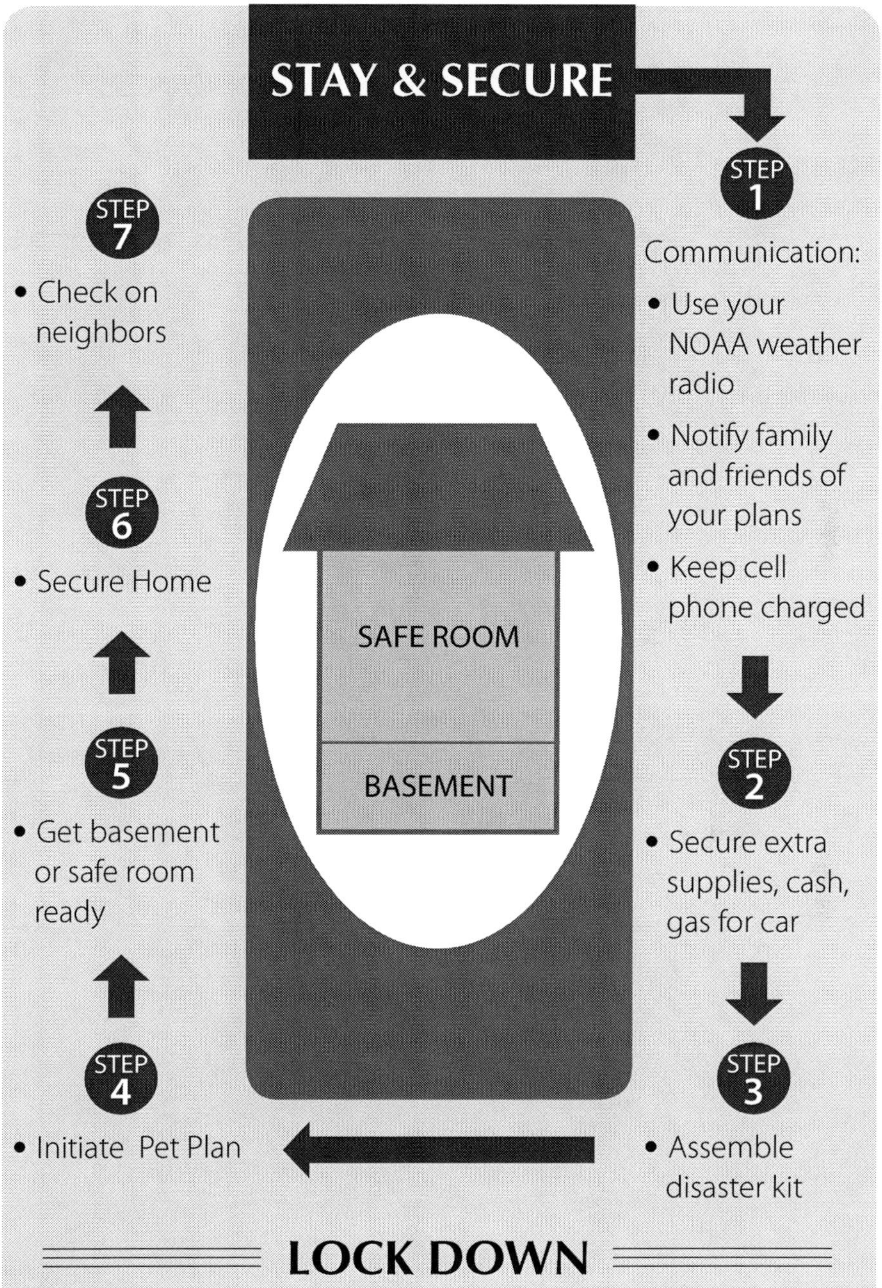
STAY & SECURE
STEP 1
Communication:
• Use your NOAA weather radio
• Notify family and friends of your plans
• Keep cell phone charged
STEP 2
• Secure extra supplies, cash, gas for car
STEP 3
• Assemble disaster kit
STEP 4
• Initiate Pet Plan
STEP 5
• Get basement or safe room ready
STEP 6
• Secure Home
STEP 7
• Check on neighbors
SAFE ROOM
BASEMENT
LOCK DOWN

Earthquake

- Drop, cover, and hold.
- If indoors, stay there.
- If outdoors, find a spot away from power lines, street lights, overpasses, trees, buildings.
- If in a vehicle, pull over at a safe location & stop.
- If in a high-rise building, do not use elevators.
- If in a coastal area, move to higher ground.
- If in a mountainous area, be alert for falling rocks, falling debris and landslides.

Hurricane Watch

(WATCH = Possible within 36 Hours)

Checklist:

- **Remember: If you wait too long and power is knocked out, you will not be able to pump gas, get groceries or cash.**

- Stay tuned to your local radio and/or television stations for information from the National Weather Service and Civil Defense regarding possible evacuations in your area.

- Check water, food and other supplies in your hurricane survival/ evacuation kit.

- Check your supply of batteries for flashlights, emergency lighting, portable radios and televisions.

- Fill cars with gasoline.

- Secure a reasonable amount of cash.

- Secure your important papers and store in waterproof containers.

- Fill empty plastic containers with water and place in freezer for "ice blocks."

- Store all outdoor furniture, canopies and loose objects. Hurricane winds will make any flying object a dangerous weapon.

Hurricane Warning

▼ (WARNING = Possible within 24 Hours or less)

Checklist:

- **Remember: If you wait too long and power is knocked out, you will not be able to pump gas, get groceries or cash.**

- Stay tuned to your local radio and/or television stations for information from the National Weather Service and Civil Defense regarding possible evacuations in your area.

- Board up all windows securely.

- Turn freezers and refrigerators to the coldest possible settings.

- Fill up bath tub and plastic containers with fresh drinking water. Plan on about one gallon of water per person per day.

- If you live in a tsunami or flood zone, prepare to evacuate when threatened by rising waters or instructed by officials

- If you have a pool, turn off electricity and add extra chlorine.

- If you do not have to evacuate, stay at home and keep off the roads.

- Make sure the whole family knows your action plan. Advise your out of state contact about your immediate plan – evacuation or stay & secure.

STAY or GO ?
Do you have your pet plan ready?

Don't Forget Your Pet Plan

Before the Disaster

- Make sure that your pets are current on their vaccinations. Pet shelters may require proof of vaccines.
- Have a current photograph.
- Keep a collar with identification on your pet and have a leash on hand to control your pet.
- Have a properly-sized pet carrier for each animal – carriers should be large enough for the animal to stand and turn around.
- Plan your evacuation strategy and don't forget your pet! Specialized pet shelters, animal control shelters, veterinary clinics and friends and relatives out of harm's way are all potential refuge for your pet during a disaster.

During the Disaster

- Animals brought to a pet shelter are required to have: proper identification collar and rabies tag, proper identification on all belongings, a carrier or cage, a leash, an ample supply of food, water and food bowls, any necessary medications, specific care instructions and trash bags for clean up.
- Bring pets indoors well in advance of a storm to keep them calm.
- Pet shelters will be filled on first come, first served basis. Call ahead and determine availability.

After the Disaster

- If pets cannot be found after a disaster, contact local animal control office and American Humane Society.

Citizen Guidance on the Homeland Security Advisory System

GREEN (Low Risk)

- Develop a family emergency plan. Share it with family and friends, and practice the plan.Visit www.Ready.gov for help creating a plan.
- Create an "Emergency Supply Kit" for your household.
- Be informed. Visit www.Ready.gov or obtain a copy of "Preparing Makes Sense, Get Ready Now" by calling 1-800-BE-READY.
- Know how to shelter-in-place and how to turn off utilities (power, gas, and water) to your home.
- Examine volunteer opportunities in your community, such as Citizen Corps, Volunteers in Police Service, Neighborhood Watch or others, and donate your time.
- Consider completing an American Red Cross first aid or CPR course, or Community Emergency Response Team (CERT) course.

BLUE (Guarded Risk)

- Complete recommended steps at level green.
- Review stored disaster supplies and replace items that are outdated.
- Be alert to suspicious activity and report it to proper authorities.

YELLOW (Elevated Risk)

- Complete recommended steps at levels green and blue.
- Ensure disaster supply kit is stocked and ready.
- Check telephone numbers in family emergency plan and update as necessary.

- Develop alternate routes to/from work or school and practice them.
- Continue to be alert for suspicious activity and report it to authorities.

ORANGE (High Risk)

- Complete recommended steps at lower levels.
- Exercise caution when traveling, pay attention to travel advisorie.
- Review your family emergency plan and make sure all family members know what to do.
- Be Patient. Expect some delays, baggage searches and restrictions at public buildings.
- Check on neighbors or others that might need assistance in an emergency.

RED (Severe Risk)

- Complete all recommended actions at lower levels.
- Listen to local emergency management officials.
- Stay tuned to TV or radio for current information/instructions.
- Be prepared to shelter-in-place or evacuate, as instructed.
- Expect traffic delays and restrictions.
- Provide volunteer services only as requested.
- Contact your school/business to determine status of work day.

Source: Homeland Security developed with input from American Red Cross

Riding On Public Transportation

- Know where the emergency exits are located.
- Be alert for suspicious individuals in the act of hiding or abandoning an object and quickly leaving the area.
- Take notice of unattended packages or objects.
- Watch out for unusual items with wires, batteries, tanks, or bottles.
- Be aware of items with a note or message attached.
- Stay away from a suspicious mist, gas, cloud, odor or seeping fluid.
- Inform the driver of any suspicious activities or objects.
- In an emergency, remain calm and stay focused.
- Listen for public announcements.
- Follow instructions given by appropriate authorities.
- Proceed in an orderly fashion.

Watch Your Back

"Gladys, come on in. The water's warm."

DRAW: Bob swimming in the ocean. A shark is behind him.

PART 4

Personal Stories

"Tell me a fact and I'll learn.
Tell me a truth and I'll believe.
But tell me a story and it will live in my heart forever."

——— *Indian Proverb*

Directory of Stories

Category	No.	Story	Page
GENERAL	S 1	The Power of Example	197
	S 2	Meet Mr. Murphy	199
	S 3	Buster Brown	200
	S 4	Father Knows Best	202
	S 5	Each Day is a Gift	203
MEDICAL	S 6	Step on It!	204
	S 7	Sirens in the Night	206
ACCIDENT	S 8	Honeymooners	208
	S 9	Too Many Cooks Spoil the Soup	209
	S 10	Kite Flyer	210
	S 11	Know Yourself	212
	S 12	The Posse	213
MISSING	S 13	Follow the Yellow Brick Road	216
	S 14	King of the Mountain	218
VICTIM	S 15	Aunt Phyllis	220
	S 16	Vacationitis	222
FIRE	S 17	Breaking News!	223
NATURAL DISASTERS	S 18	Windstorm	224
TERRORISM	S 19	Emergency Codes, Plans & Strategies	225
	S 20	Crisis Management Plan in Your School	229
	S 21	Lock Down	230
	S 22	Modern Day Pony Express	232

FROM THE HEART....

We have a family tradition during the holidays. After Christmas dinner before the opening of presents, we gather round the dinner table and share stories. Everyone has an opportunity to share an experience over the past year or talk about a favorite childhood memory. It's amazing to find out how much you don't know about your own kids or even more surprising how much your children don't know about your past lives B.C. (before children).

The following are some of our stories...

◆ *From the Heart*

THE POWER OF EXAMPLE

How many times have you been told that your kid looks just like you? Or heard the statement from friends and relatives, "Boy, doesn't Junior do everything just like Dad does?" When the kids are little, such sentiment usually elicits a smile and laughter. Sometimes, as in our family, a lot of laughter when our older son teases the younger one that he has my "walk."

Not a complimentary comparison I surmised from our younger boy's reaction.

You might argue that there isn't anything anyone can do about family similarities. It's in the genetic makeup determined by DNA and the hand you're dealt by providence. Perhaps that's true regarding the physical characteristics. More than likely it's also true about the many similarities and idiosyncrasies siblings exhibit like facial expressions. From the time the little ones can see, they're watching your every move. They learn by mimicking your actions, your language and your values.

Parents are, whether we like to acknowledge this fact or not, at different stages of our children's lives their substitute heroes, mentors, teachers, idols, or examples. So what do you suppose your children are thinking when they see you, their parents, forget to put your seatbelt on, drive 70 miles in a 30 mile zone, flip a lighted cigarette out the window in a dry summer heat, fly into road rage and cut someone off on the freeway, drink and swim, light fireworks illegally, ignore hurricane warnings and disrespect policemen, firemen, lifeguards, and anyone in authority?

At first thought you probably think I'm woefully exaggerating. On second thought, some examples might be popping up from your memory bank of some relative or someone you respected, doing exactly what I'm talking about and you decided that if they did it, maybe it was all right. Let's not put our children in such a compromising position. They need to learn to make decisions which are based on their own value judgment. Don't make the hard choices any more difficult because of the negative influences they pick up from people they love.

The power of example is not something to take lightly. Use it wisely.

MEET MR. MURPHY

Everybody has met Mr. Murphy at one time or another but they never seem to remember what he looks like. Let's try to jog your memory. It usually starts with a cup of hot coffee. You're driving to work in the morning and your cell phone rings. You reach over to grab it before the caller hangs up but you misplace the cup in the coffee holder and the coffee spills onto your new gabardine pants. The phone stops ringing. Caller ID tells you it's your boss. You have a meeting with some important clients at 9:00 a.m. at the office. It's 8:40 a.m. and you're five miles from town. You try to call your boss to let him know you're caught in traffic. Your phone battery is dead. Now you're really worried because the traffic is slowing down to a bumper to bumper snail crawl. You turn off the radio when you hear the lyrics "O What a Beautiful Morning."

Road rage starts to take over. You're determined that nothing is going to make you late. You pull up to an inch of the car in front of you so the hot rodder in the souped up Chevy can't cut in. You stare straight ahead so the little old lady in the Mercedes won't turn into your lane. Now you breathe a little easier because the red light has changed so traffic is flowing again. But just as you enter the intersection your car stalls. You try not to panic but after the third or fourth turn of the ignition you notice the gas gauge reads EMPTY. Your first thought is to remain calm. Without a second thought, you start screaming out loud and pounding the steering wheel. This can't be happening. What did you do to deserve this? The car horns are blaring from all sides and everyone is glaring and shaking their fists at you. Just as you start going down the list of whom to blame for this mess starting with your teenage son for not filling up the tank, the guy who was driving the hot rod knocks on your window. He volunteers to help push your car to the side so you're not blocking traffic. Next, the lady in the Mercedes pulls along side and tells you she's called the emergency towing service to come and help.

Relieved, you thank the good Samaritans for their help despite your behavior towards them. When you get to the service station you call your boss to explain. He tells you he called to let you know the clients couldn't make it and canceled the meeting. Don't worry, he says.

Murphy's Law declares that when one thing goes wrong, everything goes wrong. The only thing that will keep the snowball from rolling further downhill is a change of attitude. The Samaritan's kindness could have been the catalyst for changing your attitude. Now you can put a face on Mr. Murphy. He's the guy that causes you to take unnecessary risks which put you, your family and others in danger. When he knocks, don't open the door.

◆ *From the Heart*

BUSTER BROWN

Our good friend, Kathy, who worked at our veterinarian's office warned us about keeping a third puppy. "Three dogs are too many," she said. We ignored her. Considering ourselves tried and true "dog lovers", we decided to add "Thunder" to our menagerie along with his sire "Buster Brown" and our original Golden Retriever, Koa, Buster's dame. From the beginning, we knew it was a mistake. Thunder never learned the behavioral skills which were instinctual in the first two generations. He pestered Koa until she decided to put him in his place with a snap and a growl. Buster was more patient and could only be described as an indulgent parent. Tough love was not his style.

One day Thunder wandered off from our three acre parcel in the country. He didn't respond to our calls and whistles so I started off down the gravel road leading to the neighbor's papaya farm in search of the lost pup. It wasn't long before I saw the mischievous six month old running with his tail between his legs whimpering in distress chased by a big black Lab, growling and biting on his heels.

Thunder barely made it to the top of the hill. I'll never forget the image of what happened next. There, at the entrance to our driveway, stood Buster Brown, long golden hair flowing in the breeze, his huge chest puffed out defiantly. Thunder smartly dove in between his father's legs cowering in terror from his flight. The black lab stopped dead in it's tracks. The two male dogs faced each other. I froze too, not knowing what to expect and thinking wild thoughts about how I was going to separate two 100 pound plus dogs. But then something wonderful happened. It was as if there was some unspoken understanding between two warriors. One who was staking out his turf, the other who was protecting his son, prodigal though he might be. The lab seemed to nod in respect, turned and trotted back down the road. Thunder, emboldened by his father's presence, barked repeatedly at his retreating attacker as if he had scared him away all by himself.

Good parents will always rise to the occasion to protect their young. It's instinctual. It's a shared value. It's just common sense.

◆ *From the Heart*

FATHER KNOWS BEST

There's been a rash of reunions lately for the most popular of our family television series beginning with "Father Knows Best," "My Three Sons, "The Donna Reed Show," "The Dick Van Dyke Show," "Lucille Ball and Desi Arnaz," "Family Ties," and the lovable "Huxtables."

Perhaps it's a symbolic return to simpler times – a time when we all took the most basic freedoms for granted. The freedom from fear. There were no threats of terrorists, no scare of kidnappings, no danger of shootings in our schools or even the inkling of child molestation in our communities.

Truth be told, the most common theme espoused after the Twin Towers tragedy in New York has been a return to basics. More people are focused on the importance of family and reevaluating their core values and beliefs. More people are making life changing decisions that will affect their communities and society for years to come. The pendulum has swung back full tilt - with some exceptions.

One glaring change has been the realization that Father did not always know best. Actually, the subtle message even back in the 50's was that Mother really knew better. But in order to keep her man happy she let him think he was smarter. Well, gentlemen, the secret is out. Long ago it was scientifically proven that women as a rule are physically dominant in the sense that they can stand more pain, they outlast men in endurance on the track or in the bedroom, and they outlive men on average by 10 years.

Now research shows that women are surpassing men in the academic arena including the long held subjects men aspired to such as mathematics and architecture. Statistics show that the education gap is widening even further today as more than 60% of college graduates are what we use to refer to as the weaker sex. Father Knows Best indeed.

◆ *From the Heart*

EACH DAY IS A GIFT

Before the introduction of year round schooling, summer vacation seemed like an eternity. Remember your small kid time from June through August and up to Labor Day when you had absolutely nothing you had to do. Leisure time with friends at the beach, shopping trips downtown, barbecues on the grill with corn on the cob fresh from the farm.

Baseball games, roller skating and touch football in the park were part of the daily routine. Movies at the drive-in was a favorite especially when we could sneak friends in by hiding in the trunk of the car. Dances to the tunes on the jukebox were followed by snacks at the pizza parlor and strolls through the park hand in hand with your latest crush. Time stands still for the young at heart.

And then many of us grew up. Some of us grew old. Some grew bitter. We no longer looked forward to summer vacation anymore. What was the sense? It just meant hot and humid weather. We still had the daily grind of working that 9-5 job. As adults our time is consumed by worries about job security, bills, the economy, taxes and by the fear of crime and disasters natural or otherwise. We used to worry about our kids smoking or getting pregnant. Now we worry about our kids getting kidnapped or involved with crack cocaine or heroine. We worry about Aids, we worry about planes dropping from the sky. We worry about shootings in school. We worry ourselves to death without any joy.

There is another way to look at 9-11, If you know anyone who has had an operation because of a heart attack or other life threatening disease. They will tell you that they believe they are blest because they got a second chance. They will tell you that they consider each day a gift. A gift not to be wasted on worry and complaints. It is a gift to cherish and to enjoy. Grab a hold of that second chance. Be a child again, if even for a day. How do you do that? Find a playmate and take a summer vacation. Time will stand still again.

◆ ***From the Heart***

STEP ON IT !

"Doctor, she's not the type to complain," I said, " I found her doubled over in the bedroom and had to force her to come in to see you today." My sister-in-law held her sides with both arms, the anguish on her face from the pain clearly visible.

The doctor felt Lenny's pulse. Her face was pale and she shivered uncontrollably. He asked her to stick her tongue out while he looked down her throat with a mini flashlight. He used the same light to peer into her eyes and ears. Then he flipped his stethoscope over his head and placed the steely round part first on her chest, then moved it up and down her back.

"Take a deep breath," he intoned. "Exhale slowly." "Again."

"Once more," he spoke more determined this time. By the look on his face I could tell this was not good.

"I can't do anything for her here," the doctor confessed, "I think you should take her to the hospital."

Wilcox hospital was 30 miles away and could take more than an hour drive . I didn't think Lenny could take it that much longer. Anxiously I escorted Lenny out into the waiting room where her husband Paul was pacing back and forth. "We've got to get her to the hospital right away," I said to Paul. "All right," he replied in his Australian accent, "I'm just a bit away in the car park."

"With traffic the way it is right now I think we should get the ambulance," I urged. We both looked at Lenny and saw her nodding her head feverishly.

After a tense ride weaving through the one lane highway from Princeville to town, Paul riding with Lenny in the ambulance and me barely keeping up with the grace of God by ignoring red lights and irate drivers, we arrived at the emergency room.

Breathing easier now that Lenny was surrounded by doctors and nurses, Paul and I huddled together in the hallway waiting expectantly for word of her condition. The emergency doctor appeared in less than ten minutes.

"Your wife has an appendicitis," he said with urgency, "if we don't operate immediately, her appendix might burst and the poison could be fatal." Paul, frightened by the seriousness of the situation, readily agreed that the doctor should do whatever was necessary.

The good news is that the operation went smoothly and Lenny recovered in no time. Sometimes when we get together we all laugh about the wild ride in the ambulance. But deep down we also realize that it might have been just a matter of minutes that kept us from suffering a tragedy. When in doubt, be proactive and step on it!

◆ *From the Heart*

SIRENS IN THE NIGHT

We had just settled down in bed with a good book for the night when Buster and Koa, our golden retrievers, started acting up. Both dogs stood at the bedroom door whining, tails held high indicating they were not going to lie back down on their 6 foot round LL Bean pillows.

"You don't suppose they have to go out again,? "I asked my wife innocently, hoping she would volunteer to get up, get dressed and let the dogs out the back door for potty duty. The look on her face said it all. Guess I forgot it was my turn - her look definitely confirmed it. So I was the one who got up, put on my robe and dutifully trudged out to the kitchen, dogs lapping on my heels.

As soon as I opened the back door, I heard the sirens in the distance. No doubt that was what perked up the dogs' ears. The shrill sound approached closer so I took a quick scan of the rooftops around the neighborhood to check for smoke. There was none that I could detect so hopefully there was no fire. Must be a medical emergency I thought. As if reading my mind, a flashing EMS ambulance pulled up two doors down from us at our neighbors the Andersons. Less than a minute later, a fire truck led by two police cars and a bright red rescue Humvee converged at the same time and parked one behind the other blocking my driveway. The rescue squad rushed into the house and I noticed lights coming on upstairs. The EMS guys in white brought out a stretcher and medical supply bag and entered the house quickly.

My wife suddenly appeared at my side fully dressed. Out of concern more than curiosity we wondered if there was anything we could do to help. Some of the other neighbors gathered together for the same purpose. We soon learned that Ellen Anderson's mom who was visiting from Seattle had a heart attack. The rescue team had stabilized her and eased her breathing with an oxygen mask . They planned on taking her to the hospital as soon as possible. Ellen and her father would be driving down to the hospital together. We all volunteered to take care of the cats and see to it that the house was secure. Ellen's father praised the emergency crew saying he had never seen such a quick response after calling 911. We were lucky, he said, that we lived where we did.

Sometimes we take so many of our blessings for granted. Maybe we should go out of our way to thank the medical emergency personnel who respond in the middle of the night to help people in need. Maybe we could make their job a little easier by doing simple things like clearly illuminating the street address so they can find the location better at night. Maybe...

HONEYMOONERS

"Did you hear about Bob?" The staff at the hotel huddled in groups speaking in low tones - almost whispers - as if they were afraid to speak out loud about the tragic accident. Bob was one of their own. The memory of his wedding celebration just two days ago was still fresh in everyone's mind and it was hard to face the fact that the prospect of his recovery from his head and spinal injury was grim. It happened over the weekend. Bob and his new bride, Debbie, were celebrating their honeymoon at a swank hotel. In the spirit of adventure, they had found a secret swimming hole which the guidebooks made sure was no longer secret. Unfortunately, it was low tide when Bob dove in head first. He never thought anything like this could happen to him. He was young, strong, invincible. His wife waited for him to come up. Minutes passed. Finally he floated up to the surface face in the water. Bob will never walk again.

Choices. Decisions. In a spilt second we sometimes throw caution to the wind and come up face down.

TOO MANY COOKS SPOIL THE SOUP

We were back at the old hometown visiting Grandpa and Gramma. The boys slept most of the way down in the back of our conversion van for the two and a half- hour ride. So as you can imagine, as soon as we pulled into the driveway, bursting with excess energy, they started throwing snowballs and chasing Bowser, Uncle Tom's Gordon Setter out in the back yard covered knee deep in snow.

Gramma welcomed us with a tray spread full of Christmas cookies, eagle brand pie and a pot of fresh coffee on the stove. After checking on the two boys from the large picture windows in the family room, we eagerly sat down to the table in the kitchen to sample some of the freshly baked ginger bread cookies. That's when Grandpa answered the phone. It was Aunt Pearl, calling from California.

"Sally (that's our boys'Gramma)," she asked, "Did you hear about what happened to Sandy's little girl?," Without waiting for an answer, Aunt Pearl launched into how she had heard that Jackie who was about 8 years old at the time had come running into the kitchen chasing her sister Jamie. Jamie tripped and fell, bumping into the stove. Well, wouldn't you know it, Aunt Pearl continued, the frying pan was full of bacon grease and the hot grease spilled all over that little girl's face and arms. Poor little thing. She'll have scars all over herself for life, don't you think so, Sally?

We heard later that little Jamie did go to the hospital and was treated for severe burns. Turns out the story had a few twists to the version we first heard from Aunt Pearl. Apparently the two girls were playing and Jamie did run into the kitchen right into her Mom who was taking a pot of spaghetti noodles off the stove. Thankfully, it was hot water that spilled and not boiling grease so Jamie's burns were not as severe as first imagined. Whoever thought that cooking can be dangerous. Sometimes those old expressions handed down to us by the older generations have more than a ring of wisdom to it. Too many cooks in the kitchen can spoil the occasion.

◆ *From the Heart*

KITE FLYER

I wasn't home at the time but I could picture Mike hobbling up the driveway on his crutches with that sheepish grin on his face. Knowing he could easily gain my wife's sympathy, I'm sure he embellished his version of the accident for all it was worth. He had gone kite surfing for the first time. His partner, Loren, must have been new to the sport also because neither of them followed the instructions for safety. With the wind blowing at 20 knots, Mike was only partway strapped in when Loren unexpectedfly let loose the sail. Up, up and away sailed Mike, he actually hummed the tune as he recounted the incident, over the beachfront, across the highway, headed straight for the mountain. Fortunately, he landed in a grove of trees which broke his fall. But he would have hung there in the trees with internal bleeding and broken bones for who knows how long if the Emergency Medical Service crew didn't arrive quickly to rush him to the hospital. "Lucky to be alive," he chimed and proudly showed off his cast and described the pins set in his broken ankle. The ambulance and rescue crew arrived out of nowhere Mike reminisced. "Lucky to be alive," he kept repeating.

It was just the very next week when my wife and I attended the Red Cross training to renew our first aid certification when Mike's story was brought up again. We were discussing the value of cell phones in an emergency. Diane, a nursing instructor at the community college, mentioned how she was driving on the North Shore several months before and thought she barged in on an E.T. movie sequel being filmed. A man, waving hysterically at her, flew in front of her SUV windshield. Jamming on her brakes, she looked out her window and noticed that he was dangling from a huge kite and was about to crash land. She called 911 immediately and reported the location, the situation and the emergency. Diane finished her story by saying the 911 operator told her they had already received several calls about the flying kite man but thanked her anyway for her trouble.

Cell phones. One example of the best use of technology. "Lucky to be alive," as Mike would say.

KNOW YOURSELF

It was one of those perfect early summer days when the sky is crystal blue, not a cloud in sight, no humidity, no mosquitoes buzzing, a cool breeze to keep the sun from heating up. I was on the tennis court at Izatys Lodge on the south shore of Lake Mille Lacs with two of the owners' sons, Rob and Steve, practicing our high lobs. Rob hit a perfect shot that bounced on the edge of the back court line ricocheting off a pebble that sent the ball spinning erratically. Determined to prove my mastery at the game, I back pedaled as fast as I could then turned quickly to display a fancy backhand return. Unfortunately, I misjudged the distance to the green colored metal fence which surrounded the court and crashed into it at full speed. Dazed, I fell over in a heap, blood oozing from a cut over my right eye.

The Dubbs boys rushed to my side immediately, debating whether or not to call an ambulance. After recovering my senses, I convinced them that I didn't need an ambulance but agreed that my wife could drive me to the emergency room at the hospital just to make sure the cut wouldn't get infected. They half carried me over to the swimming pool where my wife was lounging with some friends. She jumped up at the sight of the blood on my shirt. Concerned, she rushed to my side to find out what happened. She squeezed my hand but when she looked under the handkerchief I had pressed to the cut, she fainted. Now the Dubbs boys really had their hands full. I'm holding on to my wife and they're still trying to keep me from falling.

Before I go any further, let me say that everything turns out fine. One of our good friends drove me and my wife to the hospital where the doctor on call stitched up the cut and gave me a shot for the infection. Later, we tried to analyze what went wrong. Yes, jumping up so quickly in the hot sun and running over to me could have been a cause for her fainting spell. Or possibly the worry about the depth of the wound and the sight of blood. Ultimately we concluded what many people have come to realize. When the person who is injured is one of your loved ones, someone who can usually handle an emergency, may become helpless in this type of situation.

We shared this story at a Red Cross First Aid renewal course. The instructor, who was certified as a nurse and first responder, related how the same thing happened to her not long ago. Her teenage son was mowing the lawn and while pushing the mower uphill, stepped in a hole and tripped. Still holding on to the mower, the blades cut off the toes on his right foot. In the ensuing scramble to rush him to the emergency room, she realized she was too distraught to drive so she asked a neighbor to assist her.

Know yourself. If you know your strengths and your weaknesses, you will be much better prepared for any emergency.

◆ *From the Heart*

THE POSSE

The flashing red lights were unmistakable in my rear view mirror. I pulled over to the side of the road to let the police cars pass. Must be an accident up ahead, I thought. I was never so far wrong. I was the culprit. The police were after me. Before I could put my seat belt on thinking that I was up against the "Click It or Ticket" vice squad, another car with flashing red lights pulled up from the opposite direction and stopped right in front of me.

One of the men in blue approached my SUV with steely eyes and a grim look. He gave me the standard one - two command – let me see your driver's license and registration.

"What seems to be the problem, officer," I say in my most ingratiating tone. Nothing but an icy stare greets my stab at polite conversation.

More reinforcements soon arrived. There are now four police cars with flashing lights in the middle of a town that is so neighborly there's not even a bank or saloon. Why so many cops? If you don't want to know, don't ask. I can tell the officer who now approaches me is the DUI specialist. He addresses me professionally. "Mr...so and so, we received a report that you were driving erratically so we pulled you over to find out what's really going on."

"What report?", I ask surprised.

"Several people called in your license numbers on their cell phones," he says casually, like that's a common occurrence. "They say you were weaving on the road."

Great, I think to myself, the techie posse is in town. I try as hard as I can to remember the car that was following me.

"Would you mind taking a little test?," the officer says.

Sensing my chance at redemption, I eagerly jump out of the car. " No problem," I volunteer.

First he makes me follow his index finger up and down, left to right, right to left. Next, he checks how cross eyed I am by pointing his finger closer and closer between my eyes until he almost touches my nose. Finally, he asks me to stand straight with my hands at my side, head tilted back into the sun with my eyes closed. Under his breath, he tells me to let him know when 30 seconds are up. Just to make sure, I ask, "Do you want me to count out loud?"

"No!," he says.

I don't say anything until I count one thousand one up to one thousand thirty silently. Then I announce confidently in a loud voice, "30 seconds up!"

All four badges standing in a semicircle are now confronting me with their problem. This is the situation. It's 12:30 in the day, I haven't been drinking, I'm not on any medication, I'm coherent as best as they can tell and most importantly, I easily passed the sobriety test. So, why did they get the call?

To ease their concern, I tell them the story about how I had been in an accident twenty years ago. "Have any of you ever driven in snow, on icy roads in 20 degree below zero weather?, I ask.

"No," they reply and start to listen even more attentively.

"It was early morning in Northern Minnesota, the snow plows had already dropped their salt and sand on the highway but the ice was so thick it was still slippery in spots. I was driving into town when I hit a icy spot which sent my car in a tailspin. Four cars approaching from the south hit the same slippery spot and started weaving back and forth across the highway. I missed the first car which would have hit me directly on the driver's side. The second car rammed into me head on and the force of the collision catapulted me into the ditch. The third and fourth cars managed to get back control. There's more to the story," I said, noticing that I had their interest," but to make my point quickly , ever since then, whenever there's an approaching car close to the median strip, I tend to veer to the right side of the road and then back to the middle of the lane. Maybe that's why it seems like I was weaving."

Did they believe me? Maybe. It was an answer to their problem so they thanked me for my cooperation and left me with a warning, "Drive carefully."

◆ *From the Heart*

FOLLOW THE YELLOW BRICK ROAD

Have you ever parked your car in one of those high rise multi level parking structures and got lost? Come on now, admit it. We can talk. The different levels could have been color coded or alphabetized, A,B,C,D. There were signs posted in the elevator in large letters – You are on the RED level.

No matter. You were probably engaged in a deep conversation with your friends or you were lost in thought about having to talk to your child's teacher that day or worse yet, your boss expects you for a private chit chat on Friday afternoon. Not a good sign.

Whatever the case, you are now back in the elevator with your car keys in your hands staring at the elevator buttons. Your friends are no longer with you. They have gone their separate ways. You are all alone. You feel the beads of sweat forming on your upper lip as you try as hard as you can to remember if you were on the C level or G level. Why didn't I write the alphabet code of the right level down on the time stub ? You promise yourself that this will never happen again.

How about this one? You're visiting relatives in Los Angeles so you rent a car for the duration. One day you agree to meet your sister for lunch at the Orange County Shopping Mall. You're happy that you find the Neiman Marcus store right away so you don't keep your daughter-in-law waiting as she has to get back to work by 1:30. After an enjoyable meal you shop for a while and your arms are full of packages as you head back out to the parking lot. The first wave of panic hits you as you stare out into the acres of cars lined up row upon row. Looking for the rental beige Honda that looks just like the thousands of cars stretched out in front of you is no easy task. Thankfully you remember that your National car keys has the license number of your rental. You spend 40 minutes going up and down the aisles before you find your match. Your arms are aching, the bags with your gifts are ripped, your blouse is glued to your back, your feet are aching from the blisters. This, you promise yourself, will never happen again. Sound familiar?

These are examples of being inconvenienced and probably made for some good laughter at the dinner table. The important thing is that even as inconsequential as these experiences might have been, the lesson learned could be potentially life saving. Always focus on a marker so you can find the trail. Follow the Yellow Brick Road home.

KING OF THE MOUNTAIN

All the kids gathered around Charlie in the parking lot on Sunday morning right after church. There were no community playgrounds in that section of town. Charlie was the oldest and the coolest guy in the neighborhood. And, truth be known, most of the girls thought he was the cutest boy they ever saw. Some dropped their bikes, others took off their skates and listened to what Charlie had to say.

"Who's gonna climb the mountain with me?," Charlie asked excitedly.

Silence. Charlie looked around at all the faces with their eyes cast down. Nobody looked up. Nobody volunteered. Neighborhood legend instilled fear into anyone thinking about challenging the mountain.

"Chicken!," Charlie teased.

Frannie, my five year old sister, stood up. "I'll go with you, Charlie."

At the time she would have followed Charlie around the world. Not wanting to be outdone, I quickly jumped up and yanked Sam up with me. Sam was my buddy from Little League and once standing in front of the other kids, he was too ashamed to sit back down. That was it. Charlie had his team.

Before we knew it, the four of us were on the clearly marked trail half way up the mountain. Charlie, of course, took the lead with Frannie on his tail, Sam followed and I brought up the rear. We laughed and joked about how easy this was and how we were going to show all those other 'fraidy cats how dumb they were not to come along. When we got thirsty, we stopped and picked the ripest guavas off the trees and sucked out the juice spitting the seeds out. Climbing the mountain was easy we boasted.

Somehow I think the mountain spirits heard us because suddenly thick brush and bushes with thorns blocked our way. The path disappeared and we found ourselves trapped between trying to climb up a steep rocky cliff with only sharp rocks to cling to or a sheer drop into the valley. Charlie put on a brave front but everyone knew he was scared along with the rest of us. It had been hours since we started the climb. The sun was sinking fast and the shadows cast by the fast approaching rain clouds made the once welcome breezes icy cold. Panic started to set in. Frannie started to cry. That did it for Charlie. He sat down on his haunches and gave up. Sam and I learned a great lesson that day. We used our fear as motivation to start screaming at the top of our lungs. No shame involved. We took off our shirts and waved it back and forth, yelling for anybody to hear.

From our vantage point on the mountain, we could see the neighborhood parking lot. We could see a loud crowd had gathered and faintly in the distance we could hear a high pitched siren. Yes, it was a siren. Two fire trucks came pulling into the parking lot. But then again, we thought, maybe it was a fire. We didn't see any smoke. We were now faced with the dilemma. What if they couldn't see us? Frannie started to cry again. And then something unexpected happened. Our father, swinging a cane knife, cut a swath through the bushes and reached us at our promontory. Without saying a word, he picked up Frannie in his arms and carried her all the way down the mountain. He never said a word to us about the incident. No lecture, no threats. Just a look at me which said it all. I had put my baby sister at risk but he also knew I had learned a valuable lesson. I'm thankful that the mountain gave me a second chance to prove it to him.

◆ *From the Heart*

AUNT PHYLLIS

My wife reminded me again that we couldn't be late to Aunt Phyllis. She would have everything ready at 12:00 noon. Her china and silver would be set on the dining room table, the chicken and dumplings would be hot off the stove and the apple pie would be freshly baked. She reminded me at exactly the time I took the wrong turn and had to back track ten miles from Crow's Nest to Eagle Lake. Contrary to popular belief about male ineptitude, I did stop at the local liquor store to ask for directions rather than depend on following my sister-in-law, Carol, who navigates by feel not by memory or by looking at a map.

It was exactly as my wife described. The dirt driveway wound around a giant oak tree past a garden full of ripe tomatoes, corn stalks and cucumber vines. We pulled up to the white clapboard house with cedar shakes on the roof and a wooden deck overlooking the garden like a scene straight out of a Thomas Kincaid painting. Cousin Sandy stopped by for the mini-reunion with her daughter, Amy, her grand daughter, Sally and husband Tom who was feeling out of sorts with a cold. All the women adjourned to the kitchen while Uncle Ad ushered me into the living room where he let me sit in his antique country rocker. I was flipping through the family scrapbook when I felt the ground tremble.

Seeing the confusion on my face Uncle Ad nonchalantly remarked, "Don't worry. That's just the train coming through." Sure enough. Just fifty feet from the family room a long Amtrak freight train with over a hundred cars hurtled through on the county tracks heading back to Chicago.

After the last car rumbled past, we all sat up to the table and in between bites of delicious fried chicken and mash potatoes, we caught up on the family gossip. Most notable that Grampa Tom must be related to the Reagan family on his mother's side, the Wilson's from Illinois. Then came the point of the story which motivated my reminiscing to begin with. Aunt Phyllis, noticing a silver pendant around sister-in-law Carol's neck, asked her if that was her rape whistle. Carol turned wine red and said "Who would want to do that to me?" Everybody laughed while I tried to figure out if I heard right. Cousin Sandy picked up the conversation by recalling how Aunt Phyllis who managed the Hallmark store at the shopping mall was walking through the enclosed parking lot when a flasher stepped in front of her and opened his coat revealing his God given birthday suit. Aunt Phyllis immediately blurted out "Well, shame on you!" Rumor has it that the culprit was so ashamed, he never exposed his shortcomings again. Now the lesson here is not meant to imply that belittling a would be attacker works all the time. It did for Aunt Phyllis. Carol still insists she won't need a whistle. My suggestion is to take Tom Patire's "Training for Life Program."

VACATIONITIS

Something happens to people when they go on vacation. Not everybody, of course, but a lot more folks than you would imagine. Could be the result of working, saving and planning so long they're just burnt out and not functioning properly. Could be too much sun, sitting on the beach drinking too many pina coladas.

Now, don't get me wrong. I'm not talking about people staying up all hours of the night dancing or sleeping in until the sun is half way across the equator for the second time. Or lounging around with a good book all day or sipping gin and tonics while sitting on the veranda watching the sunset and looking for the green flash. I'm not talking about honeymooners and anniversary couples walking on the beach holding hands and sharing private whispers. No, I actually think that's what people should do and not only on vacation.

What I am concerned about is the total loss of common sense in usually normal, responsible people. The examples are endless so I'll just mention a few which can be seen in the newspapers regularly.

"Couple Loses Heirloom"

Rob and Cindy from New York explain what happened to the two police men who responded to their call. They parked their rental convertible in a popular spot for sunbathers but didn't want to lug everything down the winding path to the secluded beach. So they took their beach chairs, beach mats, cooler with margaritas, and towels. Oh yes, and he brought the Da Vinci Code in hardback with him in his beach bag along with the latest Grisham paperback which Cindy was reading. They left Cindy's purse with her wedding ring, an heirloom from Rob's grandmother, in the back seat of the Mustang. Rob didn't want to take a chance of losing his wallet in the sand so he left it along with all their cash and credit cards in the glove compartment. When they got back to the car after a few hours they couldn't believe that anyone would break into the car and steal everything. Maybe it could happen at home, they said, but not here in the vacation land capital of anywhere USA. The police were polite and professional and filed the report.

Sound familiar? How about these?

"Lost Hiker Found"

"Teen Stranded on Cliff Rescued"

"Woman Suffers Heatstroke"

"Life Guard saves Couple Drifting Out to Sea on Inner tube"

Bet you could write some interesting scenarios for each of these headlines. None of them would pass the common sense test.

◆ *From the Heart*

BREAKING NEWS!

We had just returned home from another night of shuttling between two basketball games. With two sons in high school, our life fell into a set pattern – quick trip home to freshen up and change into sport fan casuals, back on the highway to one gym for the JV game at 6:30, then on to the next gym for the Varsity game at 8:00. If we were lucky, we stopped for Chinese noodles, schedule permitting. Not tonight.

This particular Friday night we managed to squeak by with a win in both divisions and got home still flush from the excitement of the competition. And, also very hungry. I turned on the 10:00 news while we vacillated between BLT sandwiches or going straight to the chocolate sundaes. My ears perked up when I heard the news anchor's tone of voice change. "Just in – breaking news. Our cam crew is on location at Wilhemina Rise where two fire crews are fighting to keep the blaze of a two story house under control." The screen cut from the newscaster to the live scene.

With a sense of foreboding, I sensed that the house looked familiar yet camouflaged by the smoke and flames I hoped I was wrong. I called out to my wife who was in the kitchen to come and see, hoping I was wrong. Just as she focused on the television, the announcer added, "...the location is confirmed, the fire captain informs us that the house is at 3334 Montgomery Lane and that it appears there is no one in the house." For just a second we both froze, letting the reality sink in – it was my sister's house. She and her husband were in Maine visiting his parents – but their dog Val and Sunshine, the cat, were at home with a babysitter checking in to feed them two times a day.

Without saying a word, we quickly redressed and drove back out to town trying to keep under the speed limit as much as possible but not having much control.

We drove in silence, partly because of the shock, but mostly because of fear for the worst happening and not wanting to put that into words.

After identifying ourselves, the police let us through the blockade. The fire crews had quenched the flames out by now. Only wisps of smoke escaped from the black rubble that remained of a two story 3000 square foot house. The smell of burnt charcoal lingered in the air. Susan, the pet sitter, was there with her husband Chuck. She was in tears as she described her anguish. She had been up to the house at 5:30. She took Val out for a 30 minute walk after feeding her and Sunshine. She stayed for another 30 minutes or so playing with the two leaving them in the enclosed porch. When she heard the sirens, she and Chuck who live just a block down the street rushed down and were prevented from entering the house by the fire crew. The fire was already scorching hot and the flames leaped out through the roof. Hoping the animals had made it out safely, she and Chuck combed the neighborhood calling their names over and over. Just before we arrived, the fire captain informed them that the remains of Val and Sunshine were identified by several of his crew. Our hearts ached for the sorrow she was going through and the guilt she felt because of the responsibility she had assumed. We thought of the daunting task ahead of us, having to call my sister and her husband to give them the news. Fire, like a thief in the night, has no conscience. It can rob us of our dreams, steal our loved ones away and leave in its wake the loneliness of tragedy.

WINDSTORM

The sky turned black right after we got home from school at 4:00. The kids wanted to play outside with their toy trucks in the giant sandbox their father built for them but not trusting the ominous weather I convinced them it was a perfect afternoon for popcorn and working on a new "Lego" creation. The threatening storm clouds blew over as the wind picked up. No rain. But the gusts of wind increased which made our old wooden windows creak and the shutters which weren't fastened down flap back and forth against the siding. Every once in a while we could hear a branch smack against the side of the porch.

It happened in the middle of dinner. The loud crashing sound on our roof startled us so much we choked on our spaghetti. The boys thought it was great fun but our basset hound, Hannibal, sensed the danger and scampered under the couch for protection. We hesitantly peered out of the window and saw with dismay that the beautiful Maple tree on the north side of our house was split in two. The top half of the severed tree lay on the roof of the house. My husband was afraid the weight of the tree limb which was at least a foot and a half thick would damage the roof. We were both young and inexperienced at the time. He climbed up on the roof using the cabinets on the deck as a support and started cutting up the many branches from the split trunk. Out of curiosity I climbed up on the roof to watch. Feeling nauseous, I decided to climb back down on the deck putting my lead foot on the barbecue grill as a step ladder.

Needless to say it wasn't just the tree branches flying every which way. I found myself in the classic falling nightmare with nothing to hold onto. My husband could only stare in disbelief as I fell backward off the deck and landed luckily on the many branches with lots of stems and leaves to break my fall. Unfortunately, after a quick trip to the emergency room at the hospital, the x-rays showed a fracture in three of my ribs. That was the good news. The bad news was the doctors said they couldn't do anything other than suggest I take it easy for the next few weeks and gave me some pain pills. After that experience I quickly adopted the common sense approach to climbing on roofs. That and trying never to have a cold and runny nose. Can you imagine the havoc sneezing causes with three broken ribs?

◆ *From the Heart*

EMERGENCY CODES, PLANS & STRATEGIES

In 1983 our family made the big decision to move back home to Hawaii after spending 16 years living in the beautiful state of Minnesota. Our plan was to drive our two cars to California in order to ship the autos over by boat. We would have a mini-vacation in L.A. and by the time we flew over, the cars would be waiting for us at the dock. We already sent the dog over on a direct flight from Minneapolis to Honolulu so she was already spending time in quarantine. My wife drove her sporty convertible, I took the family four door. We let our two sons choose which car to ride in with the understanding that they would have to shift cars at every rest stop. They cut cards, loser got to ride with Dad.

The plan was to drive from northern Minnesota down to Iowa, head west through Nebraska, Colorado, turn south to New Mexico then through Arizona up to Nevada and then cross over to California. We gave ourselves two weeks and took our time stopping and staying at resorts with a golf course. Nine holes or 18 holes, it didn't matter. The strategy to overcome the boys' anxiety about moving 5000 miles away from the home they grew up in and leaving their friends behind was to let them golf in every state we drove through. It worked.

Another strategy we used was the what-do-we-do-if-we-get-separated-on-the-freeway plan. These were the days before the popularity of the cell phone. So I devised a plan that if we got split up somehow on the freeway and got lost, we would both backtrack to the closest Shell gas station and wait. Do nothing. Just wait. Of course, in a panic, neither of us followed the plan. Luckily, I decided to drive up in the middle of the freeway median strip so my wife would see me parked no matter which way she was driving back and forth on the freeway. Despite our sons' embarrassment at all the cars honking at us, we did find one another after a lot of stress and worry.

These days technology has given us an advantage. I cannot emphasize enough the importance of giving your kids a cell phone for emergency use. Of course, you don't want them racking up a $500.00 monthly bill. Give them phone cards with a 10 minute time limit. Also, while we're on the subject, perhaps you should consider a family code. Something only you and your family can share. Create an emergency evacuation plan from your house in case of fire. Arrange for a safe place or prearranged location to meet in case of a flood or other disaster and you can't get back home. Have a list of important numbers of people to contact in case of emergency.

Finally, be aware of what's going on around you at all times. I know that most of us are jaded about safety requirements. For example, when was the last time you listened to the flight attendant go through the routine of fastening your seat belt on an airplane. After 9-11, I make sure I know where the exits are and who is sitting in front, across and behind me. You should too. And, most importantly, you should know that your kids do.

◆ *From the Heart*

CRISIS MANAGEMENT PLAN IN YOUR SCHOOL

My wife's favorite Arnold Schwartznegger's film is not the "Terminator." You guessed right, it's "Kindergarten Cop." The scene depicting the utter confusion of Arnold's five year olds during a fire drill brings to mind the air raid drills we held in school way back when. Only the boomers would remember the command "Get under your desk, tuck your heads between your knees, hold both arms over your head." We snickered during the whole exercise and thought it was play time. After all, in the fifties, World War II was over and we won!

Ironically, the movie imitates life. After performing the fire drill perfectly the second time, there's absolute pandemonium when the real fire breaks out. What's even more telling, is the scene where Arnold finds two teenagers smooching in the lounge. "What are you doing here," he screams, "Didn't you hear the fire alarm?" They sheepishly answer, "We thought it was just a drill."

Post 9/11, the world has changed. It's not a drill anymore. Now instead of a fire, there's more chance of a kidnapping, a shooting, a bombing, a chemical or biological terrorist threat. Depending on where you live, the natural disasters could be earthquakes, tornadoes, hurricanes and typhoons.

Does your child's school have an emergency plan for any of these crises? All of them? Do you know the plan? Does your child? Does the teacher? Has the school run through a mock emergency plan for a lockdown or evacuation? Do you know how to communicate with the school or where to pick up your child in such an emergency? What is the process, who makes the call, who's in charge?

Lots of questions to ask. Lots of answers needed. The alternative is not pretty. And you certainly don't want to be caught in the lounge like the two teens without a clue. Ask your school principal. THIS IS NOT A DRILL!

◆ ***From the Heart***

LOCK DOWN

The sunrise was especially beautiful as I drove down south early in the morning on my way to a meeting at the Hyatt. The car radio was off so I was enjoying the silence, the dramatic yellow and orange hued clouds splashed over a robin blue sky, and the aroma of my freshly brewed coffee. Without warning, the Plymouth van in front of me slammed on the brakes. As you can imagine, this caused a chain reaction down the line as far back as I could see. Luckily there were no collisions. But there's something about the law of gravity and flying liquid. My coffee came to rest in the most conspicuous spot – you know the fly zone where if you're standing in front of a hundred people giving a presentation there is no way anyone can miss the mark. I continued on my way wondering how I could dry my pants in 30 minutes.

The rubber necking was in full force as we crept slowly by the officer who was directing traffic ahead. I turned my radio on to the official Civil Defense station. Our chatty radio personality's voice announced the following news bulletin. "Police have sealed off Blaisdell Road from County Road 2 and rerouting traffic to the bypass route so ... Wait, wait a minute. This just in now. Stand by. We have breaking news. Blaisdell is now cordoned off on both entrances. Road blocks have been stationed at the entrance from County Road 2 and also from the north entrance on Wilcox Street. All traffic is being rerouted because...well, we don't have any information as to why the road is blocked just yet. As soon as we get any new information we'll be sure to pass it on to all you folks out there. Now back to our regular programming...."

Since my wife and I are empty nesters, I didn't know about the problems the elementary and middle schools on Blaisdell went through on that morning in question. I later heard from some parents that they were trying to drop their children off for school but they were turned back by the blockade. Some parents assumed it was some accident on the road so they let their children out of the car and told them to walk up the hill a couple of blocks so they could make it to school on time. In a normal situation, you might agree that this was a common sense approach. Several of the kids were older and they were all walking together in a group. As everyone found out later, the police had an armed suspect trapped in a residence on Blaisdell Road. He escaped out of a back window and ran onto the school's campus where he was finally tackled and subdued by the SWAT team.

Can you imagine how many "what if..." scenarios played through the minds of those parents who just dropped their children off at the corner? Communication, communication, communication. I can't stress it enough. The police could have informed the public better through official radio stations or at the blockade. More appropriately, they could have made sure the school was in LOCK DOWN mode and any children in the vicinity secured. As it was, there were 37 personnel involved in capturing the one suspect. None were assigned to protecting the school. That's how the story reads. I wasn't there so I'm reporting this second hand. Do you know of any similar situations where your children could have or were exposed needlessly. Let's talk to one another. For our own safety, let's make sure someone is listening.

◆ *From the Heart*

MODERN DAY PONY EXPRESS

The cowboys who were the original stalwarts of breaking the speed record for mail delivery would be appalled at the obstacles facing today's postal service employees. In some quarters, so much precaution is necessary that it takes longer to send a package to Washington by air than it takes to send a package by boat to Paris.

Here are some personal examples of the present situation. I recently called the Copyright office in Washington D.C. and was informed that their processing of mail is now backlogged up to six weeks. My wife was asked to send a gift box of flowers to people in New York, the box was returned undeliverable because the people refused to accept a package from an unfamiliar address. I am sure you all have your own personal horror stories to tell. The original threat of anthrax, the recent mailbox bombings in the central states and the overall atmosphere of terrorism has put a delay on our expectations of quick service.

Our postal employees take their jobs seriously and given the circumstances they have to operate under, they are doing an admirable job. Here are some suggestions they offer to help keep the mail moving safely and as quickly as possible.

If you receive suspicious mail:

Don't handle it if you suspect it's contaminated

Don't shake or sniff it

Don't try to clean up powder or fluids

Isolate it by covering or placing it in a sealed plastic bag

Keep others away from the area

Wash your hands for 5-10 minutes with soap and water

Call 911 and the U.S. Postal Inspection Service

Finally, get to know your daily mail carrier by name.

SHARE YOUR OWN STORIES...

Brainstorm for a minute and think about the stories in your family that haven't been told. Just think about the wonderful memories your parents can share with your children if given the opportunity. We know what you're going to say..."Yea, but...there's just not enough time in the day. Everybody works. The kids are in school. Then after school there's sports, band, homework..." We know the drill. Been there, done that.

Just imagine the next time you have a large family gathering. After everyone is done eating dinner, the game on television is over, the kids are tuckered out from playing with the cousins all afternoon, and you're asking everyone if they want dessert and coffee. Wouldn't it be the perfect chance to mention that Uncle Charlie just retired from the fire department. Then maybe you could casually ask Uncle Charlie if he would mind telling everybody about the big fire that burned down the courthouse in town...

Everybody has a relative or friend, or knows somebody that knows somebody who was either a firefighter, policeman, FBI agent, soldier in the war....tell the kids about their courage and their accomplishments.

Other Topics: Every day heroes...

- The neighbor's dog that scared the bully away...
- Cousin Cristie who rescued the whole group of girl scouts from the flash flood while camping...
- How Uncle Jimmy helped the woman give birth to her baby in the back of a Volkswagen...
- Grandpa rescued little Bobby from choking on a piece of candy...

Quick ? Quiz

Do you know the answer to the riddle?

What can keep you warm when you need to be,
Or cold when you want to be?

What can protect you from the approaching storm,
Or whisk you away to places unknown?

What can provide you with shelter at night,
Or take you on a journey in the light of day?

What can be a safe haven for Mom and the kids,
Or serve as your office instead?

What can entertain you with music, news or sports,
Or charge your cell phone and keep your food hot?

What can be a great benefit in so many cases,
Or be the cause of pain if used in the wrong places?

ANSWER: Your Automobile.
Store your Emergency Evacuation Kit and keep an extra charger for your cell phone in the car. Purchase a converter so you can turn your cigarette lighter into a 12 volt plug in.

Watch Your Back

"GLADYS, WHAT ARE YOU WORRIED ABOUT? WE'RE IN THE BIG CITY."

PART 5
Interviews

"We should learn from the mistakes of others.
We don't have time to make them all ourselves."

—— *Groucho Marx*

Directory of Interviews

I1 DEIDRE DOWNS, Miss America, a D.A.R.E. graduate 239

I2 TOM HAMILTON, father of Bethany Hamilton, shark victim 242

I3 DOUG HART, father of Jenna Hart, missing child 244

I4 SANDY HENRY, survivor of Oklahoma tornado 249

I5 MICHAEL HINGSON, from Guide Dogs for the Blind 254

I6 KRISTINE LESPERANCE, Search & Rescue Instructor 260

I7 DOUGLAS LUDOLPH, Fire Battalion Chief, L.A. County 265

I8 CHRIS SCHWART,
Morgan Stanley - World Trade Center, New York 268

I9 DARRELL SCOTT, father of Rachel Scott, Columbine victim 272

Reflection

If you could choose anyone to interview about his/her personal experience in overcoming a disaster, who would it be?

Do you believe the interview would benefit your children?

Why not put together a list of questions and make the call?

MISS AMERICA, A D.A.R.E. GRADUATE

Deidre Downs
Miss America, D.A.R.E. Graduate

NOTE: Miss America Deidre Downs graduated Magna Cum Laude with a bachelor of arts in history from Samford University, Alabama. She was also a Rhodes Scholar finalist and an Echols Scholar at the University of Virginia where she was a member of the Virginia volleyball team which advanced to the NCAA Championship for the first time. She has spent most of the year during her travels across the country promoting awareness and community action for the cure of childhood cancer. Deidre plans to use her scholarship money from the Miss America Organization to pay for her medical studies at the University of Alabama. She has set her goal to be a pediatrician.

D.A.R.E.'s (Drug Abuse Resistance Education) primary mission is to provide children with the information and skills they need to live drug-and violence-free lives. It strives to establish positive relationships between students and law enforcement, teachers, parents and other community leaders.

911: What's the most important thing that you learned from the D.A.R.E. program ?

DD: In D.A.R.E., I learned how to resist peer pressure and make good decisions for myself that would allow me to reach my goals in school, sports, and in all areas of my life.

911: What was the most influential factor in your decision to become a pediatrician?

DD: I decided to become a doctor while I was in high school after I fractured my hip and had to undergo surgery. While in college, I volunteered at a summer camp for kids with cancer and realized that I loved working with kids.

911: Who was your role model when you were growing up in Alabama?

DD: My biggest role model has always been my mother, who taught me the importance of working hard to achieve my goals. She taught me persistence, which paid off in my sports career (I battled back from the hip injury to earn a college scholarship playing volleyball) and in pageants (I tried 5 years before finally winning the title of Miss Alabama and then Miss America).

911: How can teens help their friends who are taking drugs?

DD: First of all, you must realize that your friends need help. Tell an adult you trust - a parent, coach, teacher, etc. - so your friend can get the help he or she needs to quit drugs and start making good decisions. One of my friends in high school who also dreamed of playing college volleyball ended up on drugs and dropped out of school. If she had gotten the help she needed earlier, she might not have lost the chance to pursue her dream.

911: Can you give some tips on how to withstand negative peer pressure?

DD: Everyone has experienced peer pressure at some point and usually more than once. It's often very difficult to withstand negative peer pressure - but the key is always to remember the consequences of doing whatever it is you're being pressured to do. It might be easier (at the moment) to give in, but what are you giving up by giving in to this pressure? You are risking your future and sometimes your life by going along with the crowd. Show that you're a strong and independent person by not being afraid to say no and to abstain from a behavior (be it drugs, alcohol, or sex) that will compromise your future and your goals.

911: How do you reconcile going from being a tomboy to becoming Miss America?

DD: I have always loved playing and watching sports. It may seem like a drastic transition from sports to pageants, but I've always believed that you can accomplish anything you set out to do. The road to becoming Miss America was a challenge, but I used that challenge to my advantage, motivating myself with the thought of achieving something seemingly out of reach. Competing in the Miss America system also enabled me to earn enough scholarship money to pay for my entire college education and half of my medical school costs.

911: Who was your favorite teacher?

DD: My favorite teacher was my 12th-grade AP English teacher. She was more like a college professor in the way she challenged us to think and go beyond the basics. She truly made literature relevant to us, and even more important, she taught us to aim high - to believe that we are capable of much more than we usually think we are.

911: What are the top three things you would advise graduating seniors to do?

DD: First of all, be open to new experiences and activities in college - whether it's a different subject area in which you take classes, or a volunteer project, or a new sport. Make the most of four years of opportunities you may never have again!

Also, remember to keep academics first on your priority list. Ultimately, you're there to get a degree, and you should be especially vigilant about academics if you plan to attend graduate or professional school - admissions committees will not overlook poor grades that you received during your freshman year. So start strong and study hard throughout college.

Finally, get a good book on personal finance so you can learn how to handle money on your own, including managing your student loans, avoiding credit card debt, and planning for the future.

SHARK ATTACK

Tom Hamilton
Father of Bethany Hamilton, shark attack victim

NOTE: I ran into Tom Hamilton at Border's bookstore where his world famous surfer daughter, Bethany Hamilton, recently had a book signing. The story about her unbelievable courage in surviving the tragedy of losing her arm in a shark attack and her return to surfing just a month later has been shared with millions of people on every network and cable television station in the country. I wanted to focus on how she was rescued, what life-saving techniques were used and what we can all learn from her ordeal.

911: Tom, can you describe what happened that day? What are the reasons you believe Bethany came through this trauma so well?

TH: Well, the doctor told me that one of the reasons is that Bethany was so physically fit and also, very important, she stayed calm. You know, she went through the Junior Life Guard program and she remembered a lot of the things she learned. She stayed calm and didn't panic so she could focus on what she had to do.

911: She must have lost a lot of blood?

TH: Yes, the doctors said that she lost more than 60 % of her blood.

911: How did Holt (Alana Blanchard, Bethany's friend, and her father, were surfing together) manage to get a tourniquet on her out there in the surf?

TH: Right after the shark attack, Holt pushed Bethany's board ahead on a wave which brought her onto the reef. Right then my son who was surfing with them paddled in ahead to call 911. While they were on the reef, Holt then used his shirt to tie Bethany's shoulder. Then he paddled into shore while Bethany held onto his surf shorts with her free arm. As soon as they reached shore, he used the strap off of the surfboard to secure a tourniquet around the stump. The ambulance wasn't there yet so they debated whether they should load her up into the back of his truck and drive into town to meet the ambulance. But then they heard the sirens.

911: The ambulance crew took over then?

TH: Yes, they rushed her into the emergency room at the hospital. Ironically, I was all set to undergo surgery on my knee. They stopped everything because of the priority of her accident. They wheeled me out of the room and rushed Bethany in. I didn't know anything about what happened until that very moment.

911: You must be proud of her.

TH: She's been just a wonderful daughter – always so positive and never a complaint. Just a really good kid.

How We Got Our Daughter Home Safely

Doug Hart
A Family's Prayers are Answered

NOTE: On Saturday, May 3, 2003, at approximately 3:00 p.m., Jenna Hart was abducted from a Toys 'R' Us parking lot in Madison Heights, Michigan. She was strapped in her car seat in her grandmother's car when it was stolen by a man posing as a good Samaritan who offered to help fix a flat tire.

Jenna's abduction was Michigan's first Amber Alert following the signing of the national Amber Alert bill by President Bush in 2003. Unfortunately, far too many child abductions end in tragedy. But in Jenna's case, everyone's efforts helped save the life of an innocent child. Doug Hart shares important information on how his daughter was brought home safely.

911: How long did it take before the police issued the AMBER ALERT?

DH: The police issued the Amber Alert about an hour after the car with Jenna in it was stolen. This seemed like forever to us but in reality, it was fairly quick because it takes some time for the police to confirm that a crime has been committed and that a child is in danger.

911: Did the FBI get involved from the beginning?

DH: The FBI was not involved. The State Police were involved, and although the Amber Alert was region-wide (Michigan, Ohio, Indiana) there was no real indication that Jenna's abductor was leaving the state. One of the things that we found interesting was that the police never came to our house. Our local police did, but only to check on us and give us whatever updates they had heard. The police kept in touch but unlike on television, our house was never filled with police officers and investigators asking a lot of questions. As a post-recovery follow up, consistent follow up and strong communication with the victim's family were two of the things we encouraged the Michigan State Police to focus on in their training related to Amber Alerts.

911: How did you convince the Detroit Tigers and the Detroit Pistons to broadcast Jenna's photo and your message on Sunday?

DH: The amazing part was, we didn't, our friends and neighbors did. We live in a small community nestled in the middle of suburban Detroit, and we were (and still are) active in the community so we knew a lot of people and a lot of people knew us. Within about two hours of Jenna being taken we had dozens of people stopping by the house asking what they could do to help. One of these was another resident whom we didn't even know. He knew the Piston's general manager and offered to ask him to post the Amber Alert on their scoreboard. Ironically, Jenna was found about a half-hour before the Piston's game but they had not gotten the news so they posted her photo and the information before the game. When they got the news that she'd been found safe shortly after the game began, they announced it and there was a standing ovation (it still makes me cry as I write this). We have another friend who has connections with the Tigers organization so he was able to convince them to post the Amber Alert information. As we suggest on our web site, if your child is missing, pull out all the stops and call in every favor you can to get the word out. We were fortunate to have friends who were willing and able to do this for us.

911: Did the Greektown Casino and McDonalds tell you they couldn't post your flyers because it was a corporate policy or because it was a local management decision?

DH: Both of them stated that it was corporate policy not to post flyers. We never confirmed this but unless they had heard the news story, I can understand their skepticism.

911: What are the top three things you would advise parents to do to prevent their child from being abducted?

DH: To be honest, we're not experts at abduction prevention. Our experience is such that we are able to give people real-life suggestions for finding their child if the unthinkable should happen. However, we do have an opinion on preventing abductions. First, never, ever leave your child unattended in the car, even for a few minutes. It still amazes us how many times we see kids alone in a car while their mother/father goes into the store or gas station. We are living proof that if a thief (or worse) wants the car, they will take it, regardless of whether or not there's a child in it. Second, make sure your child stays in a group with other children, whether walking to school or going to the mall. Third, teach your child how to know and recognize potentially dangerous situations. Some call this "stranger danger" but not all danger comes from strangers so your child should be educated on what to look for and what to do.

911: What was the most important factor in your mind that helped resolve the abduction so successfully?

DH: It is my opinion that the Amber Alert combined with the resulting media coverage that was the major contributing factor to Jenna's being found so quickly. Based on her recollection of driving around a lot and the fact that 3-5 hours elapsed from the time she was abducted until the time her abductor parked the car and left her, we are fairly certain that he must have seen the Amber Alerts posted on the freeway and/or heard them on the car radio. We still don't know why he drove around so much – did he realize he was in over his head or was he trying to gather the courage to molest Jenna – but we are thankful that he never hurt or abused her.

911: How did you decide which media outlets you would contact first?

DH: Actually, it was Lee Ann's brother who suggested we contact the media and he chose our local ABC affiliate because he had a contact there. Initially, he was the one that was talking to the media because Lee Ann was a basket case and I was out driving around with a friend looking for the stolen car. But after a few hours, he suggested that I return because the media really wanted to speak to a parent. I realized that this was our opportunity to spread the word quickly so I did as many interviews as I could and gave the media as much information and as good a "story" as possible. Although the media can be vultures, we knew we needed them and they were willing to help us in whatever way they could.

911: Who was the most helpful to you and your family?

DH: Since Jenna was abducted at 3:00 on a Saturday afternoon, the television stations were the most interested in getting our story and were the most helpful. They all took a personal interest in us and our situation and were just as eager to cover our happy ending as they were to cover the Amber Alert. The newspaper had only a small blurb about the abduction buried in the first section so obviously, it was not a big news story to them.

911: Which national organization was the most beneficial to you?

DH: The National Center for Missing and Exploited Children was the organization with the best advice on what we should do. They are a valuable resource for anyone with children.

911: Was the abductor ever caught? Or is the case still open and active?

DH: To date, Jenna's abductor has not been caught. Our assumption is that the police sketch is not a close enough resemblance so he has not been identified by any friends, neighbors, co-workers, etc. The worst part about him not being caught is that we don't know why he took the car with Jenna in it and we don't know if he'll do it, or has done it, again. We know he watched my mother-in-law and Jenna get out of the car and enter the store because he flattened the tire and met them at the cash register inside the store to ask if they needed help changing the tire. Did he target them because he thought my mother-in-law looked like an easy target or because he wanted my daughter? He changed the tire and although he might have tried to take the car before Jenna was inside, he didn't. However, my mother-in-law's purse was also in the car so was he trying to get the purse and a get-away vehicle or was the purse just a bonus to go with Jenna? There are a lot of unanswered questions but it's been over two years now and the Detroit police have a lot of other more serious crimes to worry about solving so there's a good chance we will never get the answers.

OKLAHOMA CITY TORNADO

Sandy Henry
Surviver, Oklahoma City Tornado

NOTE: The legendary May 3, 1999 tornado was the first category F5 tornado recorded in Oklahoma history with winds jumping from 100 mph to almost 300 mph. The largest of 61 twisters that hit the state on that fateful Monday afternoon was called an unusual beast which left 1800 homes destroyed and over $1 billion in damage. After 90 intense minutes of terror, disaster crews found 36 people killed and a 37 mile path of destruction. Sandy Henry was caught in that monstrous storm She and her husband, David did all the right things. But sometimes even when you do all the right things, that doesn't matter. Sandy recounts some of what happened that day and shares some advice.

911: Are homes in Oklahoma built with basements?

SH: There are very few homes with basements in Oklahoma.

911: Does the county have official designated underground storm shelters?

SH: No...There are several public buildings...schools, and colleges have been designated...some of these buildings have lower levels that are below ground.

911: How far or how long would it take to get to a shelter if you lived in a trailer park?

SH: Oklahoma City encompasses a lot of area (approximately 640 square miles) some of the rural areas could be as far away as 20 to 30 minutes. Some as close as a few minutes away.

911: Where would people who live in rural areas go for safety if there were no shelters available?

SH: Good question...there are several outlying areas that have cellars...there is a danger to that as well with storm debris blocking the doorways and not being able to get out...Our city takes information about anyone having storm shelters and cellars so that folks know after a tornado that someone could be trapped inside.

911: How much advanced warning does the Emergency Alert System give?

SH: In some cases just minutes. You have to realize that some Oklahoman's are used to hearing about tornado warnings and give little credit to the alerts...they are always somewhere else...not in your backyard.

Lately, the systems have really become very high tech and warnings are being made 30–40 minutes ahead of any significant weather systems.

911: How often does your school conduct practice drills for tornadoes?

SH: 2 times a year.

911: Fire?

SH: 2 times a year.

911: Terrorism?

SH: Lock down and safety – Randomly throughout the year.

911: You said the disastrous tornado of May 3, 1999 looked odd and did not have a funnel. Will you describe what you saw?

SH: It was dark and very red from the dirt. It appeared to be a wall of rain... - It looked nothing like a tornado because of the size. Larger than half a mile wide...

911: The traditional recommendation for tornado survival is to shelter in place. What is your feeling about this advice after your personal experience?

SH: First of all, what I went through was not the normal (if there is such a description) tornado....what I went through was a freak of nature. It took the most precious man from my life and has forever changed my life. The intensity of the tornado leveled to the ground my home that was standing only minutes prior. Smaller tornadoes do not have the intensity that this one had and normally I would say that the traditional shelter is sufficient.

I think that you need to realize where gas lines are in your home. We were in our bathroom lower level of our home.... not thinking...the main gas lines going into our home were on the outside of that bathroom...not a smart place to be. I will never forget the horrible gas smells from the ruptured pipes. I feared that we were going to catch on fire.

I was thankful that we had pillows, long sleeved shirt, pants and tennis shoes...there was so much glass, wood and metal shards and insulation that I was at least protected from having injuries related to the debris.

911: What suggestions can you give parents who can't get to their kids in school under a lock down situation?

SH: Be knowledgeable of the school's emergency plans, Have a plan for knowing where in an emergency...you would meet and what hospital location you would meet at if you have any way of making those kinds of decisions.

You have to trust school staff that is trained to handle many types of situations and they would not do anything to harm your children.

911: We appreciate your help and hope that we haven't been a burden because of the painful memories involved when you think about that day.

SH: Certainly you have not been a burden at all. I hope that your project is a successful one and that parents can find peace of mind with having the knowledge to make good decisions in an emergency situation.

911: I'm sure your comments will contribute much to our goal of encouraging people to have emergency plans ready for any situation. With your permission, we'd like to include a brief background about you and your family.

SH: I have lived in Oklahoma City my entire life. I graduated from college in 1984 with a Bachelor's of Science Degree from Southwestern Oklahoma State University. I married my sweetheart David Henry in Aug. 1984. He worked for Oklahoma Gas and Electric as a Material Management Coordinator (he loved his job! And was very good at it!) He had one class left and would have graduated with a degree from University of Central Oklahoma in Edmond, Oklahoma in 1999. We loved to travel, spend time at the lake and making each other happy! May 3rd 1999, I lost my husband in the devastating tornado in South Oklahoma City.

I have worked for Oklahoma City Public Schools for 21 years. I serve as the Director of Revenue and Treasurer. I have a twin sister, Kandy (nurse), two nieces (Angela (2nd year med student) and Shelby (Senior at Westmoore High School)) and two nephews (Scott (Hair Stylist and Alex (Senior Law student)) who I adore. My Mom and Dad both live in Oklahoma City. Mom (Ginger) retired Accountant from Oklahoma University Health Science Center Library and my Dad (Tom) retired from the City of Oklahoma City serving on the fire dept for 25 plus years and serving as Fire Chief of Oklahoma City. I have two cute dogs...Angel and Charlie.

That is me in a nut shell.

TRUST AND TEAMWORK

Michael Hingson

Through the Eyes of Love - a Story of Courage

NOTE: As was the case with many Americans, Michael Hingson's life changed dramatically on September 11th. He and his Guide Dog "Roselle" were the lucky ones, however, in that they escaped the World Trade Center attack. As a WTC survivor and as a Guide Dog user, he was thrust into the international limelight. Michael and "Roselle" have since become well-known as representatives of the strength of the human/animal bond, and have been successful in communicating the power of that bond in many media interviews and public presentations. Michael has traveled with a Guide Dog for 40 years and embodies the lifelong partnerships created by Guide Dogs for the Blind and those they serve.

Prior to becoming the National Public Affairs and Donor Relations Officer for Guide Dogs for the Blind, Michael worked on the 78th floor of Tower One of the World Trade Center as the Mid Atlantic district sales manager for Quantum Data Protection Division of Irvine, Calif., overseeing a sales force and field operations.

He has a master's degree in physics from the University of California at Irvine and is a member of the National Federation of the Blind. On August 7, 1984 then Governor Martha Layne Collins of Kentucky conferred on Michael the title of Colonel in the Kentucky Militia for contributions he made to the efforts of the Kentucky Department of Rehabilitation. Michael presently lives in Novato, California, with his wife, Karen, Panama (a career change golden retriever from Guide Dogs for the blind), and two cats.

911: How did you know it was a plane crash that shook the building?

MH: I didn't know of a crash for certain until both towers fell. What happened is that when the aircraft struck our building 18 floors above us the structure tipped toward the Southwest about 20 feet. When we began our evacuation I began smelling burning jet fuel as we walked down the stairs.

What I actually smelled were vapors because there wasn't actual fuel where we were. From the odor I and others assumed we were hit by an airplane. We were never told what happened while walking down.

911: Were you advised to evacuate your building immediately?

When did you get the announcement?

MH: No. There was no question about evacuating from Tower One. I was on the 78th floor when the attack happened. The tipping of the building, the fire others saw above us, and the falling debris outside my office window were enough to get us out of there.

911: Was there anybody in charge giving directions?

MH: That was my job for those in my office. I managed the office and took charge. I focused people and got them to go to the stairs to evacuate. Once guests in my office had left then and only then did I evacuate.

911: Describe how you and Roselle got out safely?

MH: You need to understand the process and teamwork of a guide dog team. Roselle's job is to insure we walk safely. She does not know where I want her to go nor would she know how to get there. If you will, she is the pilot moving us while I am the navigator who directs us. I am the team leader and must stay in charge. It is my job to know the locations of the stair wells and how to get to them.

That morning I was preparing to conduct some product training sessions for a number of our resellers. None of my sales team was in the office. The only other Quantum employee in the office was David Frank from our California office who was attending the training sessions as he had account responsibility for the distributor whose resellers I was to train.

After the guests were taken to the stairs by David, he returned so we could make some final preparations before evacuating. David,

Roselle and I then went to the stairs I chose. It happened that the stair well was open and usable. If it had not been so then I would have gone to one of the other two sets of stairs.

Roselle and I walked down the stairs just as we always did. Of course, the difference is that we had to descent 78 floors instead of only a few.

At the bottom of the stairs we had to run through running sprinklers to get into the lobby of Tower one. We then were moved quickly through the lobby, through the shopping arcade on the first floor and finally outside. We made it out into the sun light at 9:45AM, one hour after the plane hit our building.

911: We heard reports that you and Roselle actually helped other people find their way to safety?

MH: You probably heard about the urban legend about a guide dog who supposedly saved over a thousand people. NOT TRUE! No guide dog nor anyone else did that. I do know that our calm presence on the stairs helped others. We did set, I'm told, an example for others to follow. At the same time, we all knew that we couldn't afford to panic. If one person lost it and started a panic most of us wouldn't have made it out.

911: Can you tell us something about how people felt when the New York firemen went running up the stairway as you were coming down?

MH: That is the most vivid memory I will keep of that day. The sight of those brave people going up without hesitation. They all stopped to ask if I needed assistance. I told them I was fine. They wanted to send someone down with me, something totally unnecessary. I told them I had a friend, David, with me and that they need not worry. I and others asked if we could help them as they were carrying all their equipment. I and others were told that the firemen would be fine and that while they went up to fight the fire our job was to descend and get out. We did our job and they died doing their job.

911: Do you have an aversion to high rises now?

MH: No. If I or any of us live in fear of living and acting just because of that attack then the bad guys will have already won. We all need to move on and find a new normal for our lives. We must not let the attacks on the WTC or London paralyze us. We're stronger and better than that.

911: How has Roselle adapted to living in California?

MH: She was born here. In January, 2002 I was offered a position at Guide Dogs for the Blind to help in its fund raising efforts. In addition, I travel the world speaking about Trust and Teamwork, moving on from 9-11, and how to deal with change.

Roselle was trained here at Guide Dogs for the Blind. She is quite happy to be here where every office has one or more dogs in attendance every day. She is doing quite well. I think she enjoys it much more than trudging through the winter snows of New Jersey and New York.

911: Any advice to parents on how to prepare their kids for any kind of emergency?

MH: Encourage kids to be alert. Develop an emergency plan for such things as earthquakes, storms, etc. Help kids to understand that change is not a bad thing. DO NOT TEACH THEM TO BE AFRAID OF LIVING.

The reason I survived is that I did all I could to prepare for any imagined emergency by taking such precautions as participating in every fire drill. I went to the WTC evacuation preparedness classes. I learned the locations of all the stairs. I did not wait for an emergency to learn what I needed to know.

Children can be taught to be prepared without teaching them to be afraid. Living in fear paralyzes the body. Living with knowledge and strength empowers us to deal with more than we ever imagined we would have to face.

911: How has 9/11 affected your life?

MH: I value my family, my wife, and relationships in general more. I have a different job. Instead of selling computer products I now sell hope, trust, courage, and faith. I do not believe I am all that unique. We all can face disaster as well as good times. I have been given the wonderful opportunity to help educate people to the fact that the "handicap" I face is not blindness. The real handicap is the fear and misconception of people about blindness. Sight is not all it is cracked up to be. You don't need to see to live and function "normally". The problem is that you, like most people, were raised believing that without your "precious site" you would not be an whole person. Not true! Blind people merely use alternative techniques to perform the same tasks as you.

For example, I use a standard right off the shelf laptop to conduct my business, including answering your email. The only difference in my computer is that it has loaded on it a special piece of software which converts information displayed on the screen into voice.

I can travel to any place and be confident that I can get around. How I do that would take more time than you have here. For that information, stay tuned for the book I have begun to write.

911: Appreciate your help. I'm sure your comments will contribute much to our goal of encouraging people to have emergency plans ready for any situation.

MH: By the way, you should know that as I travel and speak I give ALL speaking fees right back to Guide Dogs for the Blind. This is a commitment I made with God and me. It is my way to give back something to the organization which has given me guide dogs over the past forty-one years.

SEARCH AND RESCUE

Kristine Lesperance Ph.D
K-9 Search and Rescue Instructor

NOTE: Kristine G. Lesperance, PhD, is the trainer for the Hawaii State Civil Defense US&R Task Force Canine Unit. She also chairs the education committee for NASAR Canine Section and serves as a Fun-SAR Instructor, SARTech II Evaluator and Canine SARTech III Coordinator. Her dogs include Makai and Raven, both certified and Teva, who is in training.

The National Association for Search and Rescue's primary goal is to aid in the implementation of a total integrated emergency response, rescue and recovery system.

911: What qualities do you look for in a dog before you make a decision to spend the time training the dog for S&R?

KL: The number one trait I need to see in a dog is sociability. The dog needs to want to be with people, even strangers. Some dogs go through a shy stage or fear stage during puppyhood, but I need to see the dog still be willing to engage with different people. Next, I need to see strong desire to play, preferably with toys-tuggie or ball. If the dog isn't interested in playing with toys, they need to have a very strong food drive combined with the above sociability. I can work with play/sociability or food/sociability. If the dog doesn't have either, then I am probably not going to be able to figure which "paycheck" will be interesting enough to get the dog to hunt for 6 hours to find it! We tolerate no people aggression and no undue dog aggression, so we watch dogs carefully for that as well. We also need dogs who are not fearful. There are many puppy tests available that aid a trainer in determining if the pup will be bold and courageous enough to be appropriate for SAR. Unfortunately, these tests are merely guides, not guarantees, so it's always somewhat of a gamble to pick out a 7 week old pup for a SAR prospect. Sometimes, the gamble pays off and the pup grows into a terrific SAR dog, other times it doesn't and you end up with a pet.

911: How long have you been involved in S&R?

KL: Since December 1990.

911: How soon after a child or person goes missing should the Search & Rescue dog teams be called in ?

KL: ASAP. In the first county I responded with, we were often called in by the responding officer, sometimes within 10 minutes upon arriving on the scene of a missing person's report.

911: Has this been your experience working with the police? Do they request help as soon as possible?

KL: Every locality is different. In California, our county was experiencing an increase in abductions and started calling SAR out immediately. In other places, it can take up to 3-5 days for the responsible agency to call out SAR. It's very dependent on the policies and procedures of the individual agency.

911: What are the primary advantages that search dogs offer? What exactly can they do?

KL: Dogs can find the scent of the missing person. This is evidence invisible to us humans. Dogs can indicate if there is scent present. Trailing dogs can follow a distinct scent once presented with a scent article. Some trailing dogs have indicated an ability to follow a scent trail emanating from a subject riding in a vehicle. Air scent dogs can clear large areas and once they find scent on the wind, work up that scent right to a subject. They can indicate that subject's direction of travel, and they can search areas that are difficult for humans to search due to terrain, vegetation, and limited visibility, and they can often search these areas more quickly than human resources. Dogs are a resource that can aid in the search for a missing subject, but should be deployed along with other available SAR resources - they are one piece of the SAR resource "pie."

911: Describe the different searches you've been on? Where, what type of terrain, night missions etc.?

KL: My goodness, this isn't an easy question. I responded in California for over 3 years. Then I responded in Eastern WA & Northern Idaho for 9 years before moving to Hawaii. I've been on many, many missions, from the Selkirks of Idaho to the mountains on the Kona side of the Big Island. Most missions have been at night. Dogs generally need to work when it is cooler in hotter climes, and in the dead of winter in Northern Idaho, we seemed always to get called out at night! I've followed a dog through the mountains in Idaho for 6 miles, trudging through several feet of snow! I've been in all types of terrain, and all types of weather, at all times of the day and night. I can't really offer you a number for the missions I've been on, but I believe it to be in the hundreds.

911: What was your most difficult mission?

KL: That's also difficult to isolate. Probably the most difficult emotionally are those looking for children believed victims of homicide. I once looked for a friend's missing father - that was one of the most difficult because I knew him and his family.

911: What was your most satisfying mission?

KL: All of the missions where we found the subject, whether or not my dog and I made the find. Every time a searcher finds the subject, we all have taken part because we did our job. Surprisingly, even the "turn-arounds" are satisfying--the missions where we have been called out, but cancelled before arriving because the subject has been found. We count those missions in the "win column" too.

911: Which was your favorite dog?

KL: I love all my dogs. My best partner to date was my first dog, Nikki. She was an amazing German Shepherd with a never ending desire to work. She and I were very successful together. She had a find just 3 months before I had her put to sleep at 11 years of age. She lived to work and to please me.

911: What was your motivation to get involved in working with dogs in Search & Rescue Missions?

KL: I was in the Loma Prieta earthquake in the East Bay. I watched the TV coverage of folks trapped in the Cypress Structure. I desperately wanted to help, but knew they had too many untrained volunteers crawling through the rubble already. I kept my eyes and ears open, read about disaster search dogs in the elementary school library (I was a teacher) and got a puppy. Then I found a search and rescue team and realized I needed a lot of training to be a SAR volunteer and trained to become a man tracker. As the pup grew up, I found SAR dog handlers to train with, and joined a state SAR dog team.

VETERAN FIRE FIGHTER

Douglas Ludolph
Battalion Chief Los Angeles County

NOTE: After 36 years as a Los Angeles County Firefighter, Doug and his wife Cheryl are spending their retirement traveling the world. They just returned from a cruise from Orlando, Florida to Spain when we caught up with them. Next trip will be back to Hawaii to visit daughter Shelly who is working on her Doctorate in Psychology. Lori, another daughter, lives in Ventura with the Ludolph's three grandkids. Their son, Keith is close by in Laguna Hills where he works as a fraud investigator for a mortgage company. Over the years Doug has seen every type of emergency - from train wrecks to earthquakes, mudslides, floods, explosions, and wildfires. We talked to him about his wealth of experiences.

911: When responding to an emergency, what information about the situation would be most helpful to you? Special ID's? Data Book?

DL: We get all the information we can from the 911 dispatchers. We'd like to know what kind of life situation we're responding to, who is involved, how many people. When we arrive on the scene, we need to know if everyone got out, or if there is still someone trapped in the building. Of course, it would be invaluable to know if anyone has special needs.

As far as a data book, many people don't realize that we keep a notebook on residential and especially commercial buildings. We record information such as plot plans for the locations, whether they have sprinkler systems, fire doors, hazardous materials, etc.

911: What is the ratio between calls for fire situations and medical emergencies? What about drug overdose?

DL: 911 calls to our stations are mostly medical emergencies rather than reports of fire. I'd say 10-20% are for fires. Depending on the demographics, in a neighborhood with mostly retirement age people we get more medical emergencies for heart attacks, stroke or falls... in a neighborhood with a lot of young children, we probably get more calls for accidents such as broken arms from sports or tree climbing. Drugs are a big problem. We've even responded to locations where people are manufacturing drugs and cause fires.

911: What do you advise people trapped in a high rise building to do?

DL: Well, we always stress never to use the elevators, get down the fire stairwells right away. If you can't get down, then go up to the sky roof.

If you're trapped in your room, block the bottom of the hallway door with towels to keep the smoke from coming in, cover your face with a wet towel and keep low. Try to attract attention at the windows so the rescue squad knows where you are.

911: Do you believe our schools are well prepared for any emergency? Businesses? People in general?

DL: Schools already have the basics on how to react in an emergency situation. They practice fire drills and lock down so many times during the year. So the other response plans should be worked into their existing procedures. Businesses need more help. People in general need more information.

911: Are you concerned about terrorism in your neighborhood?

DL: The security of our nuclear plants is a concern but my greatest fear is a biological attack. It seems to be an easy way for the terrorists to cause havoc.

911: Do you have an AED (automatic external defibrillator) on every vehicle?

DL: When they first came out, not every vehicle was equipped with an AED but now they all are. I believe the police are also equipped with units.

911: Do you have a K 9 unit for Search & Rescue operations?

DL: We had a black lab that was trained to detect all types of hazardous materials. The fire investigators from outside of our area kept calling to borrow our K-9. It got so bad the Fire Chief had to establish some protocols and limit the amount of time outside of our own area.

911: What was your most memorable rescue?

DL: You know, I have to say you remember them all. Over the years I helped deliver a dozen babies. But there was this one train wreck, I was there with some of the engineers from the railroad and I helped an injured mother deliver her baby. The little one made it. That birth was special.

911: In the drama "Ladder 49", Joaquin Phoenix plays a Baltimore fireman who gets trapped in a blazing warehouse. John Travolta plays the Fire Chief, Mike Kennedy, engaged in a desperate race to save his life. What did you think of the portrayal?

DL: My wife and I thought it was a good movie. Many of the other firefighters we know who saw the movie thought it was pretty realistic... especially how the actors used the right equipment at the right time. Of all the movies made about firefighters, "Ladder 49" is the best.

911: What advice would you give to families to be better prepared for any type of emergency?

DL: Education is the most important thing. And, of course, practice, practice , practice.

A NEW BEGINNING

Chris Schwart
Morgan Stanley, World Trade Center

NOTE: Chris Schwart worked for Morgan Stanley, the giant financial company. His office was located on the 74th floor of the World Trade Center. As luck would have it, he decided not to go to work on September 11, 2001 and his life has taken a decidedly different path ever since.

911: Since you were born and raised in Westchester, did you always have as your goal to work on Wall Street ?

CS: I've always felt that I should make it a goal because I was fortunate enough to have the opportunity available to me. I think that if you are raised in certain parts of New York or Westchester, the financial industry is often the easiest field in which to become wealthy and successful. I think it was a goal of mine for some time, but I always had other aspirations.

911: What company did you work for and where was your office located in the World Trade Center?

CS: Morgan Stanley, WTC 2, 74th Floor.

911: When and how did you hear about the first plane crashing into the Tower?

CS: I slept with my bedroom window open the night of Monday, Sept. 10th and was woken by the roaring engines of the 767 overhead. It sounded as if the plane was going to come through my bedroom window. My room mates were also woken by the noise. They came into my room and we turned on the news.

911: Describe what people's reaction including your own was to the first plane crash. Accident? Confusion?

CS: After the first plane hit we were in a state of confusion. When the second plane hit (which many couldn't see because it came from the south) we felt like we were under attack. A fight or flight reflex kicked in and my roommates and I left the apartment, got in my car and drove like maniacs out of the city. We were one of the last vehicles to make it on the FDR before they closed it.

911: What fortunate stroke of luck kept you home at Union Square that day?

CS: I had just returned from a vacation in Hawaii and was going to start a new position at the company. I procrastinated a bit and was supposed to go in on Wednesday.

911: How many people from your office died in the tragedy of 9/11?

CS: Fortunately, I think no one from my office died. I was told Morgan Stanley had one of the highest rates of survival among the businesses at WTC. Had it been an hour later casualties would have been much worse.

911: Prior to 9/11, did your building have an evacuation plan in place? Did your company have emergency procedures included in your employee manual?

CS: I never heard anything about an evacuation plan or emergency procedures although I'm sure they must have had something around. I remember thinking about it one Friday afternoon while making my way down to the lobby; it used to take about 15-20 min to get down due to crowds and switching elevators etc.

911: Did you read the 9/11 Commission report? Any comments?

CS: I didn't read it. I've heard enough about it for now. That report won't provide me with any closure, finding Osama will.

911: Can you tell us something about the role the New York firemen and policemen performed that day?

CS: The NYC Policemen and Firemen displayed courage that I had only seen in movies. I think all New Yorkers present on that day would tell you the same and that we will never forget their bravery.

911: Did 9/11 have anything to do with your decision to move to Hawaii and start a whole new life, a new career, a new beginning?

CS: 9/11 had everything to do with my decision to move to Hawaii. Some people bought vacation homes, some bought cars or whatever other pleasantries they had been afraid to buy, but the bottom line is that most people made some kind of life change and mine was to follow my dreams and build a life in paradise.

911: Thank you Chris. One last favor, would you like to give us some background about your family?

CS: I was born on January 4th, 1980 in New York and studied Psychology and Finance at LaSalle University in Philadelphia, PA. I have always had a huge interest in Psychology and for many years wanted to pursue a field in it.

I have one sister, Jennifer, who was born on November 20th 1982, currently living in LA with my mother, studying costume design at FIDM . We remain in close contact. My father lives in Atlanta, GA and he and his wife work for Delta Airlines. I still visit NYC very often as I have many good friends there. My grandmother, Dora Joyce, with whom I am very close, still resides in New York. My grandfather, Gunther Schwart, lives in Port Richey, Florida. He is somewhat of an idol of mine and I'm told we share many of the same traits. I currently reside in Waikiki with my wife Barbara where together we own and operate Sundance, a swimwear store at the Aloha Tower Marketplace. Barbara was visiting the States from Budapest, Hungary when I met her in February 2002 and we decided together to move to Hawaii shortly thereafter.

RACHEL'S CHALLENGE

Darrell Scott

Father of Rachel Scott, Columbine School victim

NOTE: Rachel Scott was one of the students killed at Columbine High School on April 20, 1999. Darrell Scott, Rachel's dad, is the founder of **Rachel's Challenge**, one of the most life-changing school programs in America. He has appeared on Oprah, Larry King Live, CNN, Today Show, O'Reilly Factor, Hannity and Colmes, Good Morning America, ABC, CBS, NBC, MSNBC, C-SPAN and has been featured in Time, and quoted in Newsweek, New York Times and many other periodicals.

Darrell has authored or co-authored 5 books including the best seller, **Rachel's Tears**, and meets with politicians and educators regularly concerning issues of violence. Here are some of the thoughts he shared with us.

911: How does your program Rachel's Challenge enable students to bring about positive change in their schools?

DS: Rachel's Challenge (RC) brings a change to the climate and culture of a school in a positive manner. Most educational goals are for head knowledge and instruction. Rachel's Challenge is a program that goes straight to the heart of the students. It is based on the life and writings of Rachel Joy Scott, the first person to be killed in the shootings at Columbine High School on April 20, 1999. Rachel wrote an essay a month before she died, challenging her reader(s) to start a chain reaction of kindness and compassion. Examples from her life are shared with students in a full school assembly. They hear the story of a young girl who was new at Columbine and had just lost her mother, and how Rachel reached out to her on her first day of school there. They hear about a young man who was handicapped, mentally and physically, who was being teased when Rachel stepped in on his behalf. Two years after Rachel's death, this young man told Rachel's sister, Dana, that Rachel was the reason he didn't commit suicide. They hear about how Rachel wrote that her life would be short, but that it would impact millions of peoples' hearts. They read that she believed she would die as a homicide victim. They see how her life has impacted President Clinton, President Bush, Elton John, the Dave Matthews Band, pro athletes, the strongest weightlifter in America, Miss America 2003, and even Chuck Norris, who dedicated his autobiography to her memory.

After a powerful assembly, the speaker conducts a 1 hour training session with a cross culture of the student and adult leadership of the school (approximately 50 people).

A banner is left in the cafeteria that proclaims, "I Accept Rachel's Challenge". There is room for every student to sign the 10' by

3′ banner.

That evening, there is a community event with students, parents, teachers, and community leaders.

We hear from schools a year later who continue to see the impact on kids and the way they treat each other. A climate of kindness and compassion is Rachel's legacy.

911: In your meetings with educators throughout the country, what seems to be their primary strategy in curbing school violence?

DS: The primary strategy of educators I meet with seems to be the typical approach of a combination of encouraging kids to report any unusual activities or threats, along with metal detectors, more guards, etc. Our strategy is to help them see what positive role models, influences, and inspiration can do to make a difference.

911: Do you feel that administrators are better prepared in dealing with students who may be emotionally challenged?

DS: I'm not sure I can be helpful with this question. I am flown in as a keynote speaker to a lot of educational conferences and speak in a lot of schools, but I'm the speaker, not the one being spoken too. I think a lot of good programs and initiatives have sprung up since Columbine.

911: Some schools are resorting to metal detectors and security guards. What's your feeling about that approach?

DS: My feeling about metal detectors and armed guards is that it is a perimeter attempt of a solution, but doesn't strike at the real issues. I have been in schools where there is a metal detector at the entrance but none at the gym door or exit door. If two kids worked together it would be so simple to bypass the system, by one going through the detector and opening the back door to his friend with all the weapons! Kids who are determined to bring destruction usually will. It is the heart prevention that needs more focus.

911: What about computer generated ID's that can track students?

DS: We are giving more and more signals to kids that we don't trust them and Computerized I.D.'s that track students would certainly seem intrusive to them. Again, a student determined to wreck havoc won't be stopped by that.

911: It was reported that communication between the different agencies was a big problem at Columbine just as it was on 9/11 because radios were set on different frequencies. Do you see any improvement and better cooperation now?

DS: I am not qualified to speak on this issue, although I am told that communications between agencies has improved greatly as a result of Columbine.

911: What are the top three things that you would advise parents to do about protecting their kids against school violence?

DS: Difficult question, because it isn't necessarily their kid that initiates the violence. I would tell parents to work on open communication with their children, so that if their child ever did need to tell them about a threat, he or she would. Second, I would encourage them to treat everyone with kindness. Violence feeds on bullying, being ignored, and loneliness. Kind people help diffuse that situation. 3rd, I would tell them to encourage their child to develop a relationship with a teacher they can trust to turn to if anything ever happened at school.

911: What national organization was most beneficial to you and your family after the terrible tragedy?

DS: I am not aware of any national organization that was beneficial to our family. Our local friends and neighbors were.

911: You've spoken to millions of people around the world and appeared on every major television program. What's the key message that you share in your talks ?

DS: The key message I share is to look for the little things that make a big difference in the lives of others. A simple compliment, a smile, a "thank you" are just as easy to hand out as a put-down, a sarcastic remark, or a frown.

911: What is your number one recommendation to a school principal about preventing school violence?

DS: If school principals would do what Terry Thompson, a principal in Indianapolis, IN did, and focused on character programs and inspirational programs like Rachel's Challenge or Positively for Kids, or Ethics USA, and then build their own school culture around that outside program it would bring the level of violence way down. This has been proven over and over. The antidote to violence is kindness. Create a caring atmosphere and you will develop people who care. There must be as much emphasis on the heart as their is on the head.

Watch Your Back

"Gladys, hurry up the light's yellow already."

DRAW: Bob stepping into crosswalk with head turned, not noticing speeding bus..

fyi 2

Chocolate, apple seeds and cherry pits along with common plants like aloe vera, asparagus fern, rubber plant and hibiscus can cause severe reactions in puppies and adult dogs.

fyi 3

The highest tsunami ever reported was 1,719 ft. (524m) high and occurred along the fjord-like Lituya Bay, Alaska, USA, on July 9, 1958. It was caused by a giant landslip and moved at a terrifying 100 mph (160km/h). This wave was so high it would easily have swamped the world tallest building.

— *Guinness World Records 2004*

PART 6

Questions, Quizzes and Likely Scenarios

"When you learn, teach. When you get, give."

——— *Maya Angelou*

Directory of Questions, Quizzes and Likely Scenarios

Difficult Questions - Valuable Answers

Ten Questions Parents Should Ask Themselves? 281
20 Questions Parents Should Ask? 281

Sample Quizzes to Prepare for the Real Challenge

Safety Puzzle 283
Signs & Symbols 284
True or False 286
When to Call 911 ? 287
When to evacuate ? 288
Observation Skills 289

What if... Could be Just a Heartbeat Away

Likely Scenarios at Home 290
Likely Scenarios at School 292
Likely Scenarios at Play 294
Likely Scenarios at Work 296
Likely Scenarios on Vacation 298
Likely Scenarios on the Road 300

Note: If you can imagine it, you can predict it. If you can predict it, you can prevent it.

Keeping Score is Difficult

"The names of the patients whose lives we save can never be known. Our contribution will be what did not happen to them. And, though they are unknown, we will know that mothers and fathers are at graduations and weddings they would have missed, and that grandchildren will know grandparents they might never have known, and holidays will be taken, and work completed, and books read, and symphonies heard, and gardens tended that, without our work, would never have been."

——— *Donald M. Berwick, MD, MPP*
President, Institute for Healthcare Improvement

DIFFICULT QUESTIONS — VALUABLE ANSWERS

Ten Questions Parents should ask themselves?

1. Do we know where our kids are this very second?
2. Do our children know how to get a hold of us?
3. Do we know the friends our children play with regularly?
4. Do our children know whom to call in an emergency?
5. Does the school know how to contact us in an emergency?
6. Do we have a safe home environment?
7. Did we teach our children how to protect themselves?
8. Do we have a family communication plan?
9. Does everyone in the family know where to go in an emergency?
10. Do our young children have a family pass word?

20 Questions Parents Should Ask?

FIRE FIGHTERS:

1. If you were responding to an emergency at my home, what information would be most helpful to you?
2. How long would it take you to get from the fire station to my house?
3. What do you advise people trapped in a high rise building to do?
4. Do you believe our local schools are well prepared for any emergency?
5. Do you have an AED on every vehicle?

POLICE:

6. Are there any drug dealers in our neighborhood?
7. How soon do you call in the K-9 corps to look for a missing person?
8. What is your recommendation for people who are sexually as-

saulted, fight back or do whatever the attacker demands?

9. Have you ever used the AMBER alert for missing children?
10. If we see anything suspicious in the neighborhood, what do you suggest we do?

CIVIL DEFENSE:

11. Regarding voluntary versus mandatory evacuation during a hurricane- what happens if people don't cooperate?
12. What's the best thing to do in a storm – leave the windows open a little so the pressure doesn't build or seal everything tight?
13. Does our state give tax incentives to homeowners to build "safe rooms" for emergency shelter?
14. What is the worst case scenario for a terrorist attack in our city?
15. Do we have enough shelters for everyone in case of a disaster?

MEDICAL FIRST RESPONDERS (EMS):

16. When someone calls in an emergency, what information is necessary?
17. Is bumper to bumper traffic causing a problem in responding on time?
18. Do you find more people ignoring the siren and flashing lights?
19. Is your service available for 911 on a 24 hour/ 7 days a week basis?
20. If the caller is cut off, can you trace the 911 call?

SAMPLE QUIZZES TO PREPARE FOR THE REAL CHALLENGE

SAFETY PUZZLE

ACROSS

1. Stop, Drop & __________.
4. This is used for light and you should never drape anything over the top of it.
8. Your smoke alarm needs these changed at least twice a year.
10. This is hot, dark, loud and scary.

DOWN

2. This sounds a warning.
3. This is your job if fire breaks out.
5. You and your family need to practice your fire escape __________.
6. Keep these out of reach from children.
7. These plug into the wall and kids should never touch them.
9. A burning house fills with this and it makes it hard to breathe or see.

ANSWER ACROSS: 1. roll 4. lamp 8. batteries 10. fire
DOWN: 2. alarm 3. escape 5. plan 6. matches 7. cords 9. smoke

SIGNS & SYMBOLS

Different colors on signs mean different things:

- RED signs mean stop or use extreme caution.
- YELLOW signs with black lettering are used to warn you to slow down because you are coming to a dangerous area.
- ORANGE signs with black lettering are used in areas where road construction is under way.
- GREEN signs mean go. GREEN signs with white lettering show directions and distances to different places.
- BLUE signs give information, like the route to a hospital or emergency shelter.

Identify the colors:

1. ________	11. ________	21. ________
2. ________	12. ________	22. ________
3. ________	13. ________	23. ________
4. ________	14. ________	24. ________
5. ________	15. ________	25. ________
6. ________	16. ________	26. ________
7. ________	17. ________	27. ________
8. ________	18. ________	28. ________
9. ________	19. ________	
10. ________	20. ________	

1
STOP
2
3
DETOUR
4
LANE ENDS
MERGE
LEFT
5
6
7
8
BIKE ROUTE
9
YOUR TOWN 2 →
← MY TOWN 5
10
YIELD
11
RAMP
45
M. P. H.
12
13
14
15
ROAD
WORK
16
17
SHARE
THE
ROAD
18
ONE WAY
19
BLASTING
ZONE
AHEAD
20
21
R
R
22
23
24
DO NOT
ENTER
25
26
27
DEAD
END
28

TRUE OR FALSE

T F

1. Scientists can predict earthquakes.
2. A sunstroke is not life-threatening.
3. You can only get a sunburn on really hot days.
4. Areas near lakes, rivers, and mountains are safe from tornadoes.
5. Boats should seek protection of a bay or harbor during a tsunami.
6. Use a doorway for protection during an earthquake.
7. Stay in the home during a heat wave.
8. People caught in the open during a tornado should seek shelter under highway overpasses.
9. Tsunamis are a single wave.
10. Big earthquakes always happen in the early morning.

WHEN TO CALL 911?

YES NO

☐ ☐ 1. Your neighbor's house is on fire.

☐ ☐ 2. A senior citizen crossing the street is hit by a speeding car.

☐ ☐ 3. Your teenager scrapes his knee playing basketball.

☐ ☐ 4. Two men are trying to force a woman into their car.

☐ ☐ 5. A woman is frantically looking for her lost 4-year-old child.

☐ ☐ 6. You see a construction worker fall from the second story and he is unconscious.

☐ ☐ 7. College girls on vacation are complaining about their sunburns.

☐ ☐ 8. You see a man coming out of the bank wearing a mask and carrying a gun.

☐ ☐ 9. Your daughter hits her older brother in a tussle so that his nose is bleeding a little.

☐ ☐ 10. The car in front of you veers off suddenly and hits a telephone pole.

WHEN TO EVACUATE?

YES	NO	
☐	☐	1. An avalanche is imminent. Your cabin is in the projected path.
☐	☐	2. Civil Defense warns everyone in flood zones to leave immediately. You live on the mountain.
☐	☐	3. A tsunami warning advises everyone to move to higher ground. You are at the beach in front of the hotel.
☐	☐	4. The smoke alarm in your house goes off. There's a fire in the basement.
☐	☐	5. The trembling from the major earthquake has stopped.
☐	☐	6. A tornado has been sighted. You have a large basement.
☐	☐	7. The winter storm brings with it freezing temperatures and six inches of snow. Travel advisories have been issued.
☐	☐	8. Mount St. Helen's has erupted. Urgent warnings about the toxic gases and ash have been publicized. Your 75 year old mother has asthma.
☐	☐	9. You are at a state tournament football game. A bomb threat has been sent.
☐	☐	10. You are a first grade teacher. Your principal has issued the code for immediate lock down .

OBSERVATION SKILLS

Test your children's awareness by challenging their observation skill.

A. You have 10 seconds only to observe the following.

- How many books are on the bookshelf in the hallway?
- Identify the items in your dresser drawer.
- What colors are the houses on our block?
- How many steps are there in our house?
- Count the number of trees in the backyard.

B. You have five seconds only to observe the following:

- How many women are in the restaurant?
- Describe the car parked next to our car.
- What stores are located in the shopping mall that we can see?
- How many people are dressed in a uniform?
- Describe the man standing in front of the arcade.

WHAT IF... COULD BE JUST A HEARTBEAT AWAY

Likely Scenarios at Home

AT HOME

- **What if...**
  You walk into the kitchen and see your two year old grand child in the cabinet under the kitchen sink with an open can of Drano...

- **What if...**
  Your wife starts choking at dinner because a large piece of chicken is stuck in her throat...

- **What if...**
  Grandpa, who is visiting from out of town, has a heart attack in the middle of the night...

- **What if...**
  Relatives are visiting for a week. Their five year old keeps pulling on your old dog Buster's tail...

- **What if...**
  You smell smoke coming from your elderly mother's room. You find her asleep with her lit cigarette burning a hole in the carpet...

- **What if...**
  You have a pool dinner party at your house. One young cou-

ple brings their two and five year olds... the couple completely forgets about watching them while they mingle with the other guests...

- **What if...**

  Your husband is up on a ladder painting the second story window. The ladder slides out from under him and he falls breaking his leg...

- **What if....**

  You come home late from a movie and the front door is open. There's a noise coming from your home office...

- **What if...**

  You're shopping at Walmart during the holidays. You stop to talk to a friend and suddenly when you turn around your 3 year old grand daughter disappears...

- **What if...**

  Your daughter calls from her friend's house which is 15 miles away and wants a ride home. A tornado warning has been sounded...

Likely Scenarios at School

AT SCHOOL

- **What if...**
  You volunteer to act as a chaperon on a class trip with your daughter's second grade class. At the aquarium, one of the students is missing...

- **What if...**
  At 7:30 in the morning you are dropping your children off at their school. Police have the road blocked off so other parents are just letting their kids off at the stop sign...

- **What if...**
  Your 14 year old comes home from school shaking and scared. After much persuasion, he tells you about the bullies he overheard talking about shooting their English teacher...

- **What if...**
  There's a bomb scare at your child's school. You're supposed to pick her up at the prearranged evacuation site. You don't know where that is...

- **What if...**
  There's a rumor that there will be a gang fight after the Friday night dance...

- **What if...**

  An obnoxious parent rushes down from the bleachers and punches the soccer coach in the face...

- **What if...**

  The school principal calls you at work to inform you that your kindergartner fell off the rope swing and broke her arm...

- **What if...**

  There's a report on the 6:00 P.M. news that one of the star football players at the local high school just collapsed during practice. There's talk about the use of steroids. Your son is on the team...

- **What if...**

  Several junior high girls have accused their gym teacher of sexual abuse. Your daughter tells you in the strictest confidence that it's true...

- **What if...**

  Your teenage son has been losing weight, he's failing in all of his classes and has become irritable and mean. You've caught him stealing money from your purse and he lies constantly. The school is starting a new K-9 drug search program to stop the drug traffic...

Likely Scenarios at Play

AT PLAY

- **What if...**
  Your daughter, doing a back somersault on the trampoline, bounces off and lands on her head...

- **What if...**
  You and your husband are horseback riding in the country when a wild boar darts out of the bushes and spooks the horses. Your husband falls off and his horse runs loose...

- **What if...**
  The neighborhood kids are out on the pond frozen over for the first time this winter. One of the kids falls through a crack but luckily is rescued by the park manager. She's wet, shivering and her speech is slurred...

- **What if...**
  Your 65 year old father used to be quite a good tennis player but has been a couch potato for the past ten years. Showing off at his 50 year class reunion, he keeps playing in the 90 degree heat until he drops to the court, with muscles cramp, his skin flush and he appears dizzy and weak...

- **What if...**
  Watching your son play in his first Little League baseball game brings back fond memories until he is hit in the head by a wild

pitch...

- **What if...**
  Recently on a weekly snowmobile poker run which your club sponsors, one of the players failed to return. When the rescue team finally found him stuck in a snow mound, he was obviously suffering from frostbite...

- **What if...**
  Despite warnings from the lifeguard, several of the neighborhood kids ventured out into the water and were stung by jellyfish. One child's tongue swelled up and she began gasping for air...

- **What if...**
  Doing a fancy somersault move on a skateboard, the father trying to impress his kids landed heavily on his neck and shoulder and couldn't move...

- **What if...**
  Out on a small sailboat, two brothers are blown out to sea before they know what happened...

- **What if...**
  Riding a bike down a steep hill, you fall and break your leg so that the bone is exposed...

Likely Scenarios at Work

AT WORK

- **What if...**
  While working as a hostess at a restaurant, you hear screams coming from the kitchen where one of the culinary staff has been burned badly by a pot full of boiling grease...

- **What if...**
  You become concerned when one of your coworkers who has been despondent over his recent job evaluation is fired for too many absences....

- **What if...**
  Two of your first grade students are in the bathroom when the Principal orders an immediate lockdown for the whole school...

- **What if...**
  As the Manager on Duty (MOD) at one of the major hotel chains, you get a call from Security advising you of a violent confrontation in the parking lot...

- **What if...**
  You are a paralegal in a law firm when a distraught husband storms into the office looking for his wife's divorce lawyer...

- **What if...**
  You hear a cry for help from a parent whose child is found un-

conscious in the resort swimming pool...

- **What if...**
  Several workers in the laundry department have fainted. There is a strong smell coming from the storage area...

- **What if...**
  While you're ringing up a client's purchase, a man pushes her down and runs off with her purse...

- **What if...**
  The elderly woman greets you warmly and says she wants to make a withdrawal. She hands you a note which tells you to put all of the cash from your drawer into her large purse or the bomb she left in the grocery bag at the loan officer's desk will go off...

- **What if...**
  It's lunch time and the foot traffic downtown in the business district is busy when the police block off the streets looking for an escaped prisoner...

Likely Scenarios on Vacation

ON VACATION

- **What if...**
  Your six year old falls off the fishing dock and disappears under water...

- **What if...**
  You're out in the middle of the State Park x-country skiing with your wife and kids. You keep going around in circles and realize you are lost...

- **What if...**
  Hiking with friends up a steep trail in the mountains, you twist your ankle and can't walk on it...

- **What if...**
  Your two children find a bear cub in the woods and lure it back to your campsite with treats...

- **What if...**
  You find the swimming hole with a waterfall that the guidebook calls a hidden paradise and your friends dare you to dive off the fifteen foot cliff...

- **What if...**
  The driver of the speed boat makes a sharp turn which causes your younger sister to fall, losing her water ski. She doesn't come up...

- **What if...**

  On a family trip, the cousins want your kids to go with them hunting for snakes in the desert...

- **What if...**

  You've been kayaking all day in the hot sun and at the end of the day you feel faint, you have a fever and you start vomiting...

- **What if...**

  You take the family to Disneyland reminding everyone not to get separated. Ten minutes later, your thirteen year old daughter is missing...

- **What if...**

  Your husband has a chest pain that won't go away. You're on holiday in a foreign country and can't find someone who speaks English....

Likely Scenarios on the Road

ON THE ROAD

- **What if...**
  Your sixteen year old daughter is on the way home from a dance across town and gets a flat tire...

- **What if...**
  You get a call at 11:00 at night from a friend of your son, telling you that they were in a car accident...

- **What if...**
  You hear on the 10:00 news about a terrible train crash in Tokyo. Your parents are in Japan visiting friends and you know they planned on using the trains frequently...

- **What if...**
  Your teenage son wants to go to the baseball game with his best friend who shows up riding in a pick up truck driven by his father who is obviously intoxicated...

- **What if...**
  The airline captain announces that the plane you're on is not safe to fly long distances because of equipment problems but corporate headquarters in Chicago is still undecided...

- **What if...**
  A motorcyclist is entering the highway from the gas station and

is hit by an oncoming van...

- **What if...**

  The car in front of you is weaving back and forth almost hitting cars in the oncoming lane several times...

- **What if...**

  You are driving the speed limit but the car behind you is tailgating and honking and obviously trying to pass you...

- **What if...**

  Your whole family is in the SUV driving to school. Your husband's face turns pale white as he yells out " The brakes don't work..."

- **What if...**

  You see a speeding car go through a red light and hit an elderly woman in the sidewalk...

Watch Your Back
"GLADYS, WHO'S AFRAID OF A LITTLE RAIN?"

PART 7

Resources

"The way to get things done is not to mind who gets the credit for doing them."

——— *Benjamin Jowett*

ORGANIZATIONS THAT INSPIRE

"If you want happiness for an hour, take a nap.
If you want happiness for a day, go fishing.
If you want happiness for a year, inherit a fortune.
If you want happiness for a lifetime, help somebody.."

——— *Chinese proverb*

Whenever there's a disaster, who are you going to call? Wherever there's a crucial need for food and emergency shelter, whom do you expect to be there? The two names that immediately come to mind are the American Red Cross and the Salvation Army. We've also included on our list the American Heart Association for their extraordinary success in fighting heart disease – the number one killer.

American Red Cross: *"Together, we can save a life"*

The American Red Cross is where people mobilize to help their neighbors – across the street, across the country and across the world – in emergencies. Through almost 900 locally supported chapters, more than 15 million people gain the skills they need to prepare for and respond to emergencies. About 4 million people give blood through the Red Cross. The Red Cross keeps thousands of our military personnel connected with their families. Over one million volunteers support 35,000 staff members to help the victims of upwards of 70,000 disasters.

Salvation Army: *"Doing the Most Good"*

More than 30 million people are helped in some way by services provided by the Salvation Army. Programs dealing with disaster relief, day care centers, summer camps, services for the aging, AIDS education, medical facilities, shelters for battered women and children, family and career counseling, vocational training, correction services, substance abuse rehabilitation and more...

American Heart Association: *"Learn and Live"*

The mission of this national voluntary health agency is to reduce disability and death from cardiovascular diseases and stroke. Education is the key to a healthy lifestyle. "You Are What You Eat." "It's Not Just a Man's Disease." If you're someone who wants to get involved, find out more about how you can save lives by being a part of the cure.

Your Favorite Organization:

There are literally hundreds of organizations that need the talents and skills that every individual can bring to the table. You just have to find the right one for you. The door's open, it's your chance to knock.

COMPANIES THAT CARE

> "Never doubt that a small group of thoughtful, committed citizens can change the world... indeed, it is the only thing that ever has!."
>
> ——— *Margaret Mead*

One of the hallmarks that distinguishes one company from the rest is the extent to which that company backs up its corporate responsibility by its actions in the community it serves. Here are some examples:

Home Depot:

The Home Depot has been recognized by Fortune as the No.1 Most Admired Specialty Retailer for 2005. Last Year, the company mobilized the largest supply and volunteer effort in its history and donated more than $ 4 million to the relief and rebuilding efforts following the devastation caused by four hurricanes within a six week period. More than 2,000 associates from outside the affected areas volunteered to assist their co-workers. The company has recently formed a partnership with the American Red Cross to help communities prepare for and recover from natural disasters.

Radio Shack:

Radio Shack teamed up with the National Center for Missing & Exploited Children (NCMEC) to create StreetSentz, an education and prevention program designed to help families keep their children safer from abduction, violence, and exploitation. The company introduced a new "Family Safety Kit" online brochure which includes important safety tips for parents. In addition, Radio Shack is the largest single-source distributor of ADAM Missing Child fax posters in the country.

In January 2004, Executive Technology magazine honored Radio Shack for its involvement in and support of helping children and families stay safer.

Wal*Mart:

Local store managers are "community ambassadors" and this local connection enables the company to address hometown needs. This empowerment gave Bill Burns, the Loss Prevention Supervisor, for District 186, the ability to create a special store code to be used in the event of an abduction or attempted abduction. Dawn Lane, the Safety Team leader of a store in Crawfordsville, Indiana came up with the idea of naming it "Code Adam" in memory of young Adam Walsh who was tragically abducted from a shopping mall in Florida and found killed.

Code Adam has become one of the country's largest child

safety programs and Wal*Mart in conjunction with NCMEC has generously offered other retailers the opportunity to implement this program against child abduction. Recently, Wal*Mart has implemented a new alert called "RoadWatch -Missing Child Alert" which posts pictures directly onto the cabs of delivery trucks.

Working with the Community

Here is a partial list of other companies volunteering their time and effort to join with organizations like the American Red Cross and Salvation Army to sponsor activities to help families in communities across the country: *AT&T Corporation, The Boeing Company, The Coca-Cola Company, The Dial Corporation, Dupont Company, GE, Hewlett-Packard, Honeywell, MasterCard International, Nextel Communications, Office Depot, Target, Verizon Communications...*

VOLUNTEER GROUPS CONTRIBUTE

Citizen Corps: CERT - Fire Corps - America's Waterway Watch

The CERT Program:

The Community Emergency Response Team (CERT) program trains people to be prepared for emergency situations in their community and neighborhood. CERT members give critical support to first responders in emergencies, provide immediate assistance to victims, collect disaster information to support first responder efforts and provide that first neighborhood help in the immediate hours following a major emergency.

It's families helping themselves and neighbors helping neighbors to prepare, respond and recover in their own neighborhood.

As of December 2004, there were 1,878 CERT programs in all 50 states and U.S. territories. New Jersey leads the list with 180 programs, followed by Texas (179), California (145), Florida (135) and Massachusetts (106).

The four hurricanes (Charley, Frances, Ivan and Jeanne) that devastated Florida in a period of six weeks in 2004 caused the largest deployment of mass care volunteers to gather together in US history. More than 25,000 homes were destroyed, another 40,000 severely damaged, more than 5 million residents left without power and over 20 million tons of debris left behind. Nine million people were under evacuation orders at some point with up to 350,000 staying in 370 official shelters throughout the hurricane season onslaught. There is no doubt that the Florida CERT Teams played a valuable role in supporting their neighborhoods and communities. All the thanks and congratulations they received for their efforts are well deserved. They serve as model examples of what CERT teams can accomplish.

Fire Corps:

Since September 11th, the increasing demands placed on the fire service, not only in response to the threat of terrorism, but by the changing demands of the community, it has become increasingly harder for many fire departments to provide the services the community expects, and to do its job in an efficient manner. This is where the Fire Corps can help.

Fire Corps is a locally-driven Citizen Corps program that allows community members to offer their time and talents to their local fire departments. Fire Corps serves as a gateway to information for and about fire department programs and meets a citizen's desire to serve as well as a department's need for support.

America's Waterway Watch :

America's Waterway Watch is a new national awareness program that asks those who work, live or recreate on or near the water to be aware of suspicious activity that might indicate threats to our country's homeland security.

Citizens are urged to adopt a heightened sensitivity toward unusual events or individuals they may encounter in or around ports, docks, marinas, riversides, beaches, or communities.

Anyone observing suspicious activity is simply asked to note details and contact local law enforcement. Do not approach or challenge anyone acting in a suspicious manner. Call 877-24WATCH.

Want to Volunteer? Call 1-800-volunteer.

- National Voluntary Organizations Active in Disaster - *www.nvoad.org*
- American Red Cross - *www.redcross.org*
- Points of Light Foundation - *www.pointsoflight.org*
- United Way - *www.unitedway.org*
- Network For Good - *www.networkforgood.org*
- The Independent Sector - *www.independentsector.org*
- Volunteer Match - *www.volunteermatch.org*
- SERVEnet - *www.servenet.org*
- AmeriCorps - *www.americorps.org*
- Peace Corps - *www.peacecorps.gov*
- Senior Corps - *www.seniorcorps.org*
- Volunteer Gateway - *www.volunteergateway.org*
- Community Emergency Response Team - *www.citizencorps.gov*
- Save a Life Foundation - *www.salf.org*
- National Citizen Corps Program - *www.citizencorps.gov*
- Make a Difference Day - *www.usaweekend.com*
- National Freedom Corps Program - *www.usafreedomcorps.gov*

GOVERNMENT AGENCIES

> "A nation, as a society, forms a moral person, and every member of it is personally responsible for his society."
>
> ——— *Thomas Jefferson*

Sometimes bureaucrats forget that government exists to serve the people. On the other hand, too often we expect the person behind the counter or desk to have all the answers. We need to change our approach. How about talking to one another like neighbors trying to solve a common problem? Ask for help when you need it. Offer help without being asked.

- USA Freedom Corps - *www.usafreedomcorps.gov*
- Citizen Corps - *www.citizencorps.gov*
- Corporation for National and Community Service - *www.cns.gov*
- Federal Emergency Management Agency - *www.fema.gov*
- Department of Justice - *www.justice.gov*
- Department of Health and Human Services - *www.hhs.gov*
- Centers for Disease Control and Prevention - *www.cdc.gov*
- National Regulatory Commission - *www.nrc.gov*
- EPA - *www.epa.gov/ceppo*
- National Weather Service - *www.nws.noaa.gov*
- U.S. Geological Survey - *www.usgs.gov*
- Office of Homeland Security - *www.whitehouse.gov/homeland*
- U.S. State and Local Gateway - *www.statelocal.gov*
- Be Ready - *www.Ready.gov*
- American Public Works Association - *www.pubworks.org*

Source: Information on Organizations, Companies, Volunteer Groups and Government Agencies adapted from media reports and public announcements.

USEFUL WEBSITES

MEDICAL

- Family Doctor - *www.familydoctor.org*
- WebMD Health - *www.WebMD.com*
- Alcohol and Drug Addiction - *(800) 382-4357*
- American Academy of Family Physicians - *www.* aafp.org
- American Diabetes Association - *www.diabetes.org*
- American Heart Association - *www.americanheart.org*
- American Lung Association - *www.lungusa.org*
- American Red Cross - *www.redcross.org*
- Epilepsy Foundation - *www.epilesyfoundation.org*
- National Alzheimer's Association - *www.alz.org*
- National Cocaine Hotline - *(800) COCAINE*

ACCIDENTS

- Mothers Against Drunk Driving - *www.madd.org*
- National Safety Council - *www.nsc.org*
- National Safekids Campaign - *www.safekids.org*
- Kids in Cars - *kidsincars.org*

MISSING PERSONS

- America's Most Wanted - *www.amw.com*
- Angel Safe Foundation - *www.angelsafe.org*
- Child Shield, USA - *www.childprotectionsolutions.com*
- Code Amber - *www.codeamber.org*
- Escape School - *www.escapeschool.com*
- Find Missing Kids - *www.findmissingkids.com*
- KlaasKids Foundation - *www.klasskids.org*
- National Association of Search & Rescue - *www.nasar.org*
- National Center for Missing and Exploited Children - *www.missingkids.com*
- Neighborhood Network Advisory - *www.nnanow.com*
- Voice for the Children - *www.voice4thechildren.org*

VICTIMS

- Crime Stoppers International - *www.crimestoppers.org*
- Drug Abuse Resistance Education - *www.dare.com*
- Guardian Angels - *www.guardianangels.org*
- Internet Crimes Against Children - *www.icactraining.org*
- Mothers Against Violence in America - *www.mavia.org*
- Neighbors Who Care - *www.neighborswhocare.org*
- Recovery Inc. - *www.recovery-inc.com*
- Samaritan Counseling Center, Inc. - *www.pnc.edu*
- Security on Campus, Inc. - *www.securityoncampus.org*
- Speak Out for Stephanie - *www.sos.lawrence.com*
- Victims Remembered, Inc. - *www.vri.windberschools.org*

FAMILY RESOURCES

- Because I Love You - *www.becauseiloveyou.org*
- Center for the 4th and 5th R's - *www.cortland.edu*
- Character Education Partnership - *www.character.org*
- Character Counts! - *www.charactercounts.org*
- Ethics Resource Center - *www.ethics.org*
- Families Anonymous - *www.familiesanonymous.org*
- Giraffe Project - *www.giraffe.org*
- Institute for Global Ethics - *www.globalethics.org*
- Parenting - *www.abcparenting.com*
- Positive Coaching Alliance - *www.positivecoach.org*
- School for Ethical Education - *www.ethicsed.org*

PETS ARE ASSETS

- Guide Dogs for the Blind - *www.guidedogs.com*
- National Disaster Search Dog Foundation - *www.searchdogfoundation.org*
- Noah's Wish - *www.noahswish.org*
- The Humane Society - *www.hsus.org*

Afterword

"Our task is not to fit the blame for the past, but to fix the course for the future."

——— *John F. Kennedy*

"If you think you're too small to have an impact, try going to bed with a mosquito in the room."

——— *Anita Roddick*

"Love one another..."

——— *John 13:34*

The John Hopkins University sponsored a program entitled *The Public as an Asset, Not a Problem – A summit on leadership during bioterrorism*. Over 170 people attended, representing senior decision –makers in public health and safety – including leaders and policy makers in medicine, public health, nursing, hospital administration, disaster relief and national security. Dr. Monica Schoch-Spana, PhD gave the welcome and opening remarks in which she outlined three collective tasks for the participants:

1. To revisit our assumptions about mass responses to crisis;
2. To explore ways to meet the public health and public safety requirements of diverse publics during a bioterrorist crisis;
3. To explore ways to equip every day institutions, such as work places, schools and civic groups to help people cope with a crisis.

There are two presentations from the distinguished panelists which we have chosen to highlight because their topics focused on the role of schools in a disaster and the role of the ordinary citizen – topics which concern families the most.

Bradley D. Stein, MD, PhD is an Assistant Professor of Child Psychiatry at the University of Southern California and a Health Services Researcher for the Rand Center for Domestic and International Health Security. His talk centered on why schools are a natural place to support children and families in the community during a time of crisis.

One of Doctor Stein's main points is that the role of schools extends beyond responding to traumatic events in their own back yard. The Oklahoma City Public Schools provided support services following the bombing of the Federal Building. The schools in the Washington, D. C. area, played an important role in counseling the children and families of the communities during and after the sniper attacks. After September 11, the New York City schools provided counseling to kids and families but schools across the country were also active in supporting children and families in response to terrorism.

After any disaster, when parents are asked what was important to them, their first thoughts were "Where are my children?" and "Are they safe?"

Another point underscored by Doctor Stein is that bioterrorism presents schools with a unique set of challenges. A biological or chemical event is different than any other disaster that we are used to dealing with such as hurricanes, tornadoes, fire or even school shootings. The use of a biological weapon raises questions such as when did the attack occur, who is at most risk and also when is the danger past?

There are no easy answers but three very concrete recommendations were offered:

1. Schools should be part of the emergency management agency planning to allow input from community partners into the school's crisis plans.
2. Schools should actively participate in emergency management community drills and tabletop exercises.
3. Schools should have a seat at the table in the community's incident command structure.

Diane Lapson is the Vice President of the Independence Plaza North Tenant Association, a housing complex of 4,000 people located three city blocks from Ground Zero. Diane was surprised that she was invited to speak at the summit, we assume because she considers herself a civilian. She shouldn't have been surprised. We think what she contributed was extremely important. Here's some of what she said...

"...Suddenly above our head came the loud roaring and the huge plane speeding downtown exploded into Tower Number 2. The next hour was horrific...someone then shouted that the Pentagon was hit, and it seemed like it was the end of the world, yet my instinct was to keep helping and we'd live through this. We is the key word. Evidently, this thought is not unique to me. It was suddenly apparent that humanity's instinct for compassion is stronger than its instinct for personal survival..."

Diane goes on to add three messages which she bases on lessons learned from her experience. We are gratified to quote her words for you.

"To the citizens, organize your neighborhood now. Start a small group, a large group, we pray no more attacks come, but do it anyway. It's good to be prepared..."

"To the government and the relief agencies, knowing people will volunteer and want to help during an emergency, accept their invitation. There may be fear that volunteers will not do the job correctly, but they're going to volunteer anyway, so training is important...:"

"The police departments, and fire departments, and even the armed forces can give us some instruction. It may not seem so important – it may not have seemed so important before, but now it's really important to us..."

Diane Lapson drew her inspiration and hope from the people in her community. On September 11, 2001, she saw everyone in New York put aside their personal issues and come together for the common good. We experienced the same feeling of togetherness after Hurricane Iniki ravaged the island of Kauai. Every one of the individuals we interviewed for this book talked about the tremendous support and comfort they received not only from family and friends but also from complete strangers. The people we talked to in Florida, in Oklahoma, in Washington, D. C., in California and throughout the country - all agree that crises seem to bring out the best in people.

As Tom Hanks the Oscar-winning actor said in his address to Vassar College graduates, his daughter Elizabeth among them, one four-letter word holds the answer to most of life's problems. "Help and you will make a huge impact in the life of the street, the town, the country and our planet."

Our task is to stand ready to support one another not only in times of trouble, but in times of complacency. Whenever we lose sight of our common goal, we tend to lose the common bond which unites us and keeps us strong. Let hindsight serve as our foresight during these critical times. Our challenge is to keep looking out for each other every day. Start with your family and move progressively outward as the circle of influence illustrates. We can so we must.

We told you in the beginning that this is an interactive book, so talk back to us. If you have any suggestions on how we can enhance future editions, let us know. We'd love to hear from you.

We wish you and your family the best that life offers. Be happy!

The Lactaoens

RECOMMENDATIONS

Here is a partial list of Products and Services which we believe will help in your efforts to keep your family prepared for any emergency.

Cell Phone

Every member of the family who is able to leave the house on their own should have a cell phone.

NOAA Radio

If you don't have a radio that can operate by electricity, battery and hand crank, you should add this to your emergency kit.

AAA Emergency Road Service

This auto club service or equivalent will save you from needless worry about anyone in the family being stranded on the highway especially at night.

Training for Life *www.cdt-training.com*

Consider enrolling the entire family in this Personal Protection Program. It will do wonders for their confidence and knowledge.

Automatic External Defibrillator (AED)

Sudden Cardiac Arrest strikes 340,000 Americans each year and 80% of the time it happens in the home. The AED helps saves lives.

Safe Return and Digital Angel

Enroll Alzheimer's patients in the Safe Return program or check on the Digital Angel which is a tracking device used as a wristwatch.

Drink Safe Technology *www.drinksafetech.com*

Don't let your child become an unsuspecting victim. Find out more about the availability of the Drink Safe Test.

APPENDIX **A**

SCHOOLS

WHAT EVERY PARENT SHOULD KNOW ABOUT A SCHOOL CRISIS

- *Do you know the Emergency Evacuation –Relocation Plan for your child's school?*
  Every school has the responsibility to take preventive actions for the safety of its students and to keep parents informed. An effective Crisis Management Plan will prepare students, parents, teachers and emergency personnel for a crisis and immediately communicate details, injury reports, status updates and relocation plans.

- *Does the school plan specifically cover terrorist threats including biological, chemical, nuclear, and radiological incidents?*
  Many schools have excellent plans for fire, natural disaster and intruder scenarios. Many do not have plans for specific terrorist threats. Find out if your child's school has such a plan.

- *Are your school's Science/ Chemistry labs equipped with additional security features, protective safety gear and surveillance monitoring equipment?*
  The hazardous materials in most basic high school laboratories can cause a tremendous amount of damage in the wrong hands.

- *What kinds of supplies does your school have on hand in case of a shelter -in-place requirement?*
  Families are being advised to have survival kits on hand in case of an emergency. Shouldn't schools do the same thing?

- *Do the police and fire departments have floor plans of your child's school?*
  If you don't know, ask. You'll probably be pleased at the answers.

A
SCHOOLS

- *Are First Aid and CPR Training and Certification mandatory for all faculty and staff?*
  During a crisis, even the best laid plans go awry. Critical life-saving techniques can be crucial until emergency medical personnel arrive on the scene.

- *Is gang violence increasing in your child's school?*
  A gang is defined as a loosely organized association of three or more persons that has a common name, signs, symbols or colors, whose members engage in violent forms of behavior.

- *Who is the school's contact at the police and fire departments?*
  It helps to have a contact who is familiar with the school's lay out and personnel.

- *How often does your child's school practice the emergency plan?*
  Most schools conduct the fire drills exceptionally well because they are familiar with it. But guess which drill they practice the most?

- *How can you contribute to your child's school emergency preparedness?*
  Find out when the next meetings are scheduled and attend.

Source: Adapted from Saf-T-Net and America Prepared Campaign

A

SCHOOLS

THE TOP THREE SCHOOL DISTRICTS IN CRISIS PREPAREDNESS

Fairfax County Public Schools

Number of Schools: 241
Number of Students: 166,601

Fairfax County Public Schools is cited as a model by the Department of Homeland Security and the Department of Education . It has an exhaustive emergency plan and parents are active in the school system which is one of the reasons for their success in getting prepared.

Their plan provides response actions for terrorist emergencies such as bomb threats, chemical and biohazard emergencies, gas or odor concerns, and hostage situations.

They do walk-throughs of shelter-in-place and lockdowns with teachers and crisis teams, and they cycle through table top exercises with each of the school regions.

Fairfax has an e-mail system whereby parents can get messages from their school and the entire district in the event of an emergency. Their website makes available comprehensive preparedness information in seven different languages (Arabic, English, Farsi, Korean, Spanish, Urdu, and Vietnamese) including specific information about what parents should do in an emergency.

Montgomery Public Schools

Number of Schools: 190
Number of Students: 139,203

Montgomery County Public Schools has an exemplary multi-hazard crisis plan, and is also used as a model for other school districts.

The district requires two code red and two code blue drills a year, in addition to 10 traditional fire drills. During the D. C. area sniper crisis, all of the 190 schools in the county system received an automated telephone alert, along with an urgent e-mail message, detailing the circumstances of the emergency and the response. Montgomery Schools has put extensive information on its website about the drills, including excerpts from two vivid videos produced by Sopris West Educational Services – one on code red and one on code blue.

Montgomery does an excellent job in communication and their director of Safety and Security, Edward Clarke, works with the Parent-Teachers Associations to disseminate information about preparedness. Materials are available in Chinese, Englsih, French, Korean, Spanish and Vietnamese.

School District of Palm Beach County

Number of Schools : 193
Number of Students : 160,223

Innovative measure introduced by School District Police Department Chief James Kelly, include the requirement of photo identification badges for all teachers, students, and visitors, a GPS tracking system for county school buses, and the introduction of several violence-intervention programs.

Communication seems to be the district's forte. Palm Beach County employs the use of the Communicator, which uses 24 phone lines and can place 3,000 calls per hour. Back-up communication plans include three different radio systems: conventional, trunking and digital. Ham radios can also be used as a last resort during an emergency.

The district has also partnered with local media in order to communicate with parents. The Safe Schools department hosts an interactive television talk show for parents on the district's Education Network, which discusses various safety issues. Critical announcements about school rules and procedures are broadcast in English, Creole and Spanish.

From The America Prepared Campaign report:
"Preparedness in America's Schools: A Comprehensive look at terrorism Preparedness in America's Twenty Largest School Districts" September 2004

A

SCHOOLS

FOR ELEMENTARY SCHOOL TEACHERS

Using *911 The Red Book for Emergencies* as resource material

Social Studies/History:

- Study past disasters and compare to present day.
  For example: Chicago fire - what they did then and what they might do now (after reading 911 for emergencies)
- Discuss individual/family responsibility in particular disaster situations.
- Have a grandparent or grandparents visit the classroom and speak about the "generation gap" and or other subjects featured in 911 - **The Red Book for Emergencies**.
- Discuss the increase in safety because of technology.
  For example: the cell phone
- Field trips to fire station or police station or visits by policeman, fireman to talk with class about the importance of being prepared.
- Your class can become "pen pals" with a class in another geographical area and share "safety tips" and information on what disasters are common or different.

Reading:

- Master reading some of the stories and or interviews (from **911 -The Red Book for Emergencies**) aloud to class. Interviews may involve 2 persons reading with expression etc.
- Visit public library (or school library) - read at least 3 books on a particular topic from **911 -The Red Book for Emergencies**.
- Write a book report on one of the natural disasters covered in the book.

English:

- Write own personal stories dealing with various safety topics.
- Orally tell about own emergency situation.

A
SCHOOLS

- Do a short skit or play with 2-4 friends acting out proper emergency actions to be taken in certain situations.
- Prepare an emergency kit - for a particular disaster and orally share contents etc. with class.
- Do a research paper/report on a recent local disaster in your community.
- As a class choose a "local hero" write thank you notes to that hero.
- Create and write own "common sense" principles.
- Interview a public official. Put interview in writing.
- As a class discuss and prepare a chart entitled:
  "Why every member of the Family Needs a Cell Phone."

Art:

- Cartoons depicting emergency scenarios.
- Pencil draw, paint, pictures depicting stories.
- Create own board game or other game dealing with subjects in **911 -The Red Book for Emergencies**.
- Take pictures of a particular local disaster (or aftermath of the disaster) display with written explanations.
- Make "cup telephones" and experiment playing "telephone" to show how communication can become miscommunication.
- Each student bring a photograph of one of their heroes with a short written explanation why this person is a hero to them.
  Create a booklet entitled "Our Class Heroes."
- Have a Class discussion on favorite movie films with everyday heroes - choose a favorite as a class.
  Attend that movie at the theatre (or rent video and watch as a class.)
- Paint, Crayon or Multi Media - have students do a large wall mural dealing with any topic from **911 -The Red Book for Emergencies**.
- Students make individual collages from magazines, newspapers, catalogues using pictures of disasters, and/ or heroes deeds.

A

SCHOOLS

Math/Science:

- Create own crossword puzzle dealing with 911 information.
- Prepare statistics charts on certain disasters in own area.
- Prepare a report on the effects of fire inhalation.
- Determine how much water, food and other items a person needs in any given emergency situation.
- Prepare a chart showing natural disasters prevalent to local area - showing times of occurrence etc.
- Study natural disasters - earthquakes, tornadoes, hurricanes, etc. Talk about the causes and conditions conducive to these natural disasters.

Homework:

- Help your family complete the forms in the back of **911 -The Red Book for Emergencies**.
- Prepare a family telephone calling tree.
- Make up a friends telephone calling tree.
- Prepare a lock down strategy with family.
- Prepare evacuation plan from home with family.
- Prepare a strategy plan if members of the family are separated.
- Compile family emergency information – contact numbers etc.

Contests for Students of All Ages

ELEMENTARY

Drawing Contest

Find any of the blank cartoons in **911- The Red Book for Emergencies** and complete the drawing!

JUNIOR HIGH

Thank you...very much!

Write a thank you letter to someone who has influenced you in a positive way!

HIGH SCHOOL

Essay Contest

Write a true story about an "everyday hero" who performed an exemplary service for someone in need.

Contact us for more information on:
Prizes / Scholarships / Guidelines / Instructions
Email: **comsense@whatyourfamilyneedstoknow.com**
See our web site at **www.whatyourfamilyneedstoknow.com**

APPENDIX **B**

SAMPLE EMERGENCY KITS

First Aid Kit:

- Two pairs of latex or other sterile gloves
- Sterile dressings to stop bleeding
- Cleansing agent/soap and antibiotic towelettes to disinfect
- Antibiotic ointment to prevent infection
- Burn ointment to prevent infection
- Adhesive bandages in a variety of sizes
- Eye wash solution to flush the eyes or as general decontaminant
- Thermometer
- Prescription medications you take every day such as insulin, heart medicine and asthma inhalers
- You should periodically rotate medicines to account for expiration dates
- Prescribed medical supplies such as glucose and blood pressure monitoring equipment and supplies
- Scissors
- Tweezers
- Tube of petroleum Jelly or other lubricant
- Non-prescription drugs:
- Potassium Iodide (Read More: nuclear blast)
- Aspirin or non-aspirin pain reliever
- Anti-diarrhea medication
- Antacid (for upset stomach)
- Syrup of Ipecac (use to induce vomiting if advised by the Poison Control Center)

Emergency Car Evacuation Kit

- Battery powered radio, flashlight and extra batteries
- Blanket/sleeping bags/extra clothing
- Booster Cables, tow rope
- Fire Extinguisher (5lb.A-B.C type)

B
SAMPLE EMERGENCY KITS

- First aid kit and manual, water proof matches
- Bottled water and non-perishable high energy foods such as granola bars, raisins, and peanut butter
- Maps, compass, shovel, flares
- Windshield scraper and brush
- Tire repair kit and pump
- Knife and other tools
- Sack of sand or cat litter
- Tissue, paper towels and sanitary napkins
- Empty containers with covers

Important Family Documents
- Keep copies of important family records such as insurance policies, identification and bank account records in a waterproof, portable container

Basic Survival Kit
- Portable radio
- Flashlights and extra batteries
- Water proof matches, lighter and candles
- Manual can opener
- First aid kit, special medications, and an extra pair of eyeglasses
- Five-day supply of non-perishable, ready-to-eat food
- Ice chest and blue-ice packs
- Containers of water (minimum 2 quarts per person per day)
- Sleeping bags or blankets
- Change of clothing for each family member
- Camp stove, lantern and fuel
- Masking tape for windows
- Personal hygiene, sanitary supplies, diapers
- Bleach
- Plastic sheeting and garbage bags

- Whistle

Special Needs:

For Baby:

- Bottles, Formula, Powdered milk
- Diapers, towelettes, rash ointment
- Medications

For Adults:

- Prescription medications
- Denture needs
- Contact lenses and supplies
- Extra eye glasses

For Seniors:

- Plan how you will evacuate or signal for help
- Plan emergency procedures with home care agencies
- Tell others where you keep your emergency supplies
- Teach others how to operate necessary equipment
- Prescription medications including dosage in your supply kits
- Extra eyeglasses and hearing-aid batteries
- Extra wheelchair batteries or other special equipment
- Style and serial numbers of medical devices
- Copies of medical insurance and Medicare cards
- List of doctors and emergency contacts

For People with Disabilities:

- Create a support network to help in an emergency
- Tell these people where you deep your emergency supplies
- Give one person a key to your house or apartment
- Wear medical alert tags or bracelets

B

SAMPLE EMERGENCY KITS

- Show others how to operate your wheelchair
- Prescription medicines including dosage
- List of any allergies
- Extra eyeglasses and hearing-aid batteries
- Extra wheelchair batteries, oxygen
- List of the style and serial number of medical devices
- Medical insurance and Medicare cards
- List of doctors, relatives or friends who should be notified

For Pets:
- Proper identification
- Immunization records
- Medications
- Carrier or cage
- Collar, leash, muzzle
- Treats & toys
- Blankets & pillows
- Food & water

Flu Survival Kit:
- 3M N95 1860 mask
- Hot/Cold pack
- Alcohol-based mini-sanitizers
- Vitamin C packs
- Bottled water & nonperishable foods

GO Kits: For under your bed or under your desk at work
- Walking/running shoes
- Flashlight & extra batteries
- Face mask & Wash cloth
- Extra cash & list of emergency numbers

Source: American Red Cross and Federal Emergency Management Agency

List of Illustrations

Common Sense Perception 34
The Circles of Influence 37
The Cycle of Generations 38
The POO Factor 39
Twelve Common Sense Principles for Emergencies 44
Elements of Communication 48
Multiple Dangers/ Multiple Risks 76
The Danger of the Wild Reference Chart 102
Natural Disaster Reference Chart 126
Which Plan Should You Take 183
Evacuate 184
Stay & Secure 185
Stay or Go? 189

List of Cartoons

What's the Family Emergency Plan 45
Watch Your Back – Tsunami 70
Watch Your Back – Shark 194
Watch Your Back – Thief 236
Watch Your Back – Bus 277
Watch Your Back – Lightning 302
Watch Your Back - Bear 332

NOTE: Bob and Gladys are our favorite "Watch Your Back" characters. Bob is an example of the unaware citizen, Gladys never appears with him in these compromising situations. Our cartoonist gives us his perspective in three of the scenarios. We've included three additional scenes for you and your kids to challenge your imagination. Have fun!

Watch Your Back

"GLADYS, DON'T BE A FRAIDY CAT. NOTHING IS HIDING IN THE FOREST."

NOTES

Introduction

1. Jean O. Pasco, Times Staff Writer, *Parents Turn Grief to Action*, May 6, 2005
2. A version of a story used by Professor Robert Sternberg, professor of psychology and education at Yale University.

Communication

1. Richard Lederer, *The Revenge of Anguished English.*,St. Martin's Press 2005
2. Ethnologue www.ling.gu
3. *Intercultural differences*, www.understandfrance,org
4. James Debussy, *Treasure Cache* magazine, January 1993, page 34.
5. An anonymous quote shared with us by motivational speaker, Dore Jean Petrich.

Multiple Dangers / Multiple Risks

1. Compiled from Dr. Pippa Keech, *The Complete First Aid Handbook*, Hermes House, 2003; American Red Cross materials and interviews with American Medical Response personnel
2. Office of Juvenile Justice and Delinquency Prevention (OJJDP), *When Your Child is Missing: A Family Survival Guide*, DOJ *In Crisis*,
3. Frank Krau," Street Safe: Life in the Real World" Seminar
4. Adapted from the *Domestic Abuse Intervention Project*, Duluth, MN
5. Adapted from the *Domestic Abuse Intervention Project*, Duluth, MN
6. Adapted from Dr. Lenore Walker's *Cycle of Violence*, Denver, CO

Afterword

1. Cited from a report by the John Hopkins University's Center for Civilian Biodefense Strategies on *The Public as an Asset, Not a Problem* , 2003.

REFERENCES

American Medical Association, *Basic Disaster Life Support*, Provider Manual 2004.

American Red Cross

American School Counselor Association Guide,

Be Ready.gov

Bioterrorism Preparedness Conference 2003 and 2004, Training made possible by the Centers for Disease Control, Atlanta, Georgia and the CDC Bitoerrorism Preparedness Cooperative Agreement

Blauvelt, Peter, *Making Schools Safe for Students*, Corwin Press,1999.

CDT, *Training for Life Program*, Certification 2004

Citizen Corps.gov

Community Emergency Response Team, Certified Participant Manual, 2003

COPS, *Developing Community Groups for Homeland Security*, Department of Justice, Participant Manual, 2003

De Becker, Gavin, *Protecting the Gift*, Dell Publishing, 1999

Federal Emergency Management Agency

Fernandez, Edward, *Safefood Solutions*, Participant Manual, 2004

Keech, Dr. Pippa, *The Complete First Aid Handbook*, Hermes House, 2003

National Education Association, *Crisis Communication Guide, 2004*

National Association for Search & Rescue, *Fundamentals of Search and Rescue*, Participant Manual 2004

OSHA *Compliance & Workplace Safety*, Certificate of Continuing Education 2004

Patire, Tom *Personal Protection Handbook*, Three Rivers Press, 2003

The 9/11 Commission Report, W.W. Norton & Company, 2004

What Your Family Needs to Know

www.whatyourfamilyneedstoknow.com

ORDERED BY: ____________
VIP Number

Name ____________

Title ____________

Organization ____________

Address ____________

City, State, Zip ____________

Phone ____________ Fax ____________

E-mail Address ____________

METHOD OF PAYMENT Check one

❑ Check ❑ Money Order
(Payable to HCI-Independents Group Press)

❑ Company Purchase Order Attached
(Please mail or fax : 808-826-9319)

❑ MasterCard ❑ Visa ❑ Discover ❑ AmEx

Credit card number ____________

Expiration date ____________

Signature ____________

Please send free information on:
❑ Other Books & Products ❑ Seminars ❑ Speaking ❑ Consulting

Buy a copy for a family member.
Buy a copy for a friend.

Title	Price	Quantity	Total
911-The Red Book for Emergencies	$16.95		
Our Family Profile Forms from The Red Book	$ 7.95		
The Green Book for Financial Security Available in winter of 2006 / 10% Pre-publication Discount	$16.95		
Common Sense Principles Available in winter of 2006 / 10% Pre-publication Discount	$15.95		
		Merchandise Total	
		10% Pre-publication Discount	
		Shipping U.S. ($4.00 for first book and $2.00 for each additional product)	
		Subtotal	
		HI Residents add 4% Sales Tax	
		Total	

A charitable donation of $1.00 per book will be made to the organization of your choice.
❑ American Red Cross ❑ Salvation Army
❑ American Heart Association

THE INDEPENDENTS GROUP PRESS • *PUBLISHER AND DISTRIBUTOR OF SPECIALTY BOOKS* •
P.O. Box 3650 Lihue, HI 96766 **Phone:** 245-4604 **Toll Free:** 1(888)245-4604 **Email:** orders@whatyourfamilyneedstoknow.com

911 Seminars Workshops

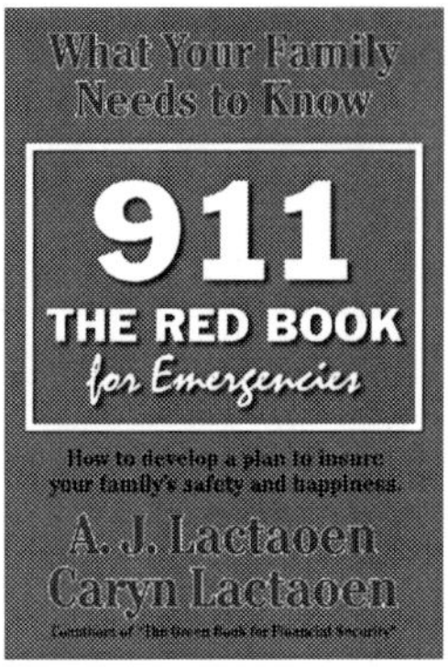

Workshops are based on the book "911-The Red Book for Emergencies."

Every 911 Workshop is divided into two parts in order to provide not only information but also the knowledge and skills needed in a crisis.

COMMON SENSE STRATEGIES 101

Basic (1 Hour and 30 minutes)

Part 1 : The instructor will help families develop plans for dealing with any type of major emergency whether it is a medical emergency, an accident, a missing person, a victim of crime, a fire, a natural disaster or an act of terrorism.

Part 2: Hands on demonstration in Adult and Child CPR, First Aid techniques and the use of an Automatic External Defibrillator

COMMON SENSE STRATEGIES 102

Plus (2 Hours and 15 minutes)

Part 1: Families will learn a life management system which empowers them to make better decisions every day.

Part 2: Hands on demonstration in all of the techniques established in Basic 101. Additional demonstrations include Infant CPR, Pet CPR, Rape Escape options and the use of a Flu Survivor Kit.

911 SEMINARS

PMB 290 P.O. Box 223300 Princeville, HI 96722
www.911seminars.com
email: workshops@911seminars.com
Phone: 808-651-4180 Toll Free: 1-800-721-8222

Back of the Book
Your Personal Family Profile

"I hear... and I forget; I see... and I remember,
I do... and I understand."

——— *Confucius*

NOTE:
This section contains forms to be filled in by you for your complete family profile.

Suggestions

Show everyone in the family where your copy of 911 -THE RED BOOK is kept. Just because it was purposely designed so it would be easy to find, doesn't mean you should bury it under a pile of mail or stack of laundry. Keep it handy and close to the phone – on your desk or on a bookcase. Make sure babysitters know where and what it contains.

Write clearly and in large block print letters so everyone can read the numbers and names easily. In an emergency, you don't want Grandfather to have to run and find his reading glasses. And for sure, you don't want the phone numbers to look confusing.

Fill out the forms completely. You don't have to do it all at once. Do it as a family and assign each one a task. Set a deadline when each person has to turn in the information or draw the diagram or do the research. No exceptions for age or seniority. Everyone should contribute.

Every form should be filled out. Of course, there are some forms that pertain only to young children and some forms only for the senior generation. If the situation doesn't apply to your family, help your friends who do need that type of information.

Little things can be so important in an emergency. For example, one parent whose little girl was kidnapped, couldn't remember the license number of the family car that was stolen and couldn't find the information right away. Precious moments were wasted as the police held off initiating the Amber Alert. Don't make the same mistake. Fill out every appropriate form now. Don't wait until it's too late.

Directory of Forms

F1 CALL 911 ... 340
F2 Family Profile ... 341
F3 Family Emergency Numbers ... 342
F4 Outside Emergency Contacts ... 343
F5 Requirements for Emergency Room ... 344
F6 Family Emergency Plan ... 345
F7 Family Phone Tree ... 346
F8 Diagram of Our House ... 347
F9 Safety Zone ... 348
F10 Danger Perimeter ... 349
F11 Grandparent's House ... 350
F12 School Evacuation Plan ... 351
F13 Work Evacuation Plan ... 352
F14 Family Secret Password ... 353
F15 Our Children's Friends ... 354
F16 Child ID Profile ... 355
F17 Physical Information ... 356
F18 Dental Chart / DNA Information ... 357
F19 Location of Emergency Supplies ... 358
F20 Our Family First Aid Kit ... 359
F21 Our Family Evacuation Kit ... 360
F22 Our Family Survival Kit ... 361
F23 Family Automobiles and Transportation ... 362
F24 Family Travel Itinerary ... 363
F25 Emergency in Foreign Country ... 364

Reflection

If a mandatory evacuation was issued and you could take only one thing with you, what would it be?

Do you know where it is?

CALL 911

DO NOT HANG UP
UNTIL OPERATOR SAYS IT'S O.K.

Be ready with the following information:

1. Your name and your contact number?

____

2. Address and/or location? Landmark?

____

3. The type of emergency or injury?

____

4. How many people injured? Ages? Gender?

____

Answer any other questions
the 911 Operator asks.

FAMILY PROFILE

LIVING IN THIS RESIDENCE ARE:

☐ WORKING ADULTS

Name ______________________________

____

Special Needs ______________________________

☐ CHILDREN

Name ______________________________

____

____

Special Needs ______________________________

☐ SENIOR

Name ______________________________

____

Special Needs ______________________________

☐ PETS

Dog ______________________________

Cat ______________________________

Bird ______________________________

Other ______________________________

FAMILY EMERGENCY NUMBERS

PARENTS

	Cell	Work	Pager
____	____	____	____
	Cell	Work	Pager
____	____	____	____

CHILDREN

	Cell	School	Work
____	____	____	____
	Cell	School	Work
____	____	____	____

GRANDPARENTS

	Cell	Home	Work
____	____	____	____
	Cell	Home	Work
____	____	____	____

RELATIVE OR NEIGHBOR

	Cell	Home	Work
____	____	____	____

OUT OF STATE CONTACT

	Cell	Home	Work
____	____	____	____

OUTSIDE EMERGENCY CONTACTS

Fire Department	Don't Stall. CALL **911**
Police Department	
Ambulance	

Poison Control Center	
Crisis Hotline	

	NAME	TELEPHONE	LOCATION
Hospital			
Doctor			
Pediatrician			
Dentist			
Pharmacist			
Veterinarian			
Child's School			
Parent's Workplace			
Electric Co.			
Gas Co.			
Water Co.			
Insurance Agent			
Landlord			

REQUIREMENTS FOR EMERGENCY ROOM

Caregiver ______________________________

Symptoms ______________________________

Medications ______________________________

____

Allergies ______________________________

____

Medical Conditions ______________________________

____

Prior Surgeries ______________________________

____

Family Physician ______________________________

Specialists ______________________________

Insurance Policy ______________________________

Representative ______________________________

NOTE: Attach copy of living will

FAMILY EMERGENCY PLAN

Plan A When everyone is together

OPTION 1 ____________________

OPTION 2 ____________________

OPTION 3 ____________________

Plan B When everyone is separated

OPTION 1 ____________________

OPTION 2 ____________________

OPTION 3 ____________________

Plan C When communication is not possible

OPTION 1 ____________________

OPTION 2 ____________________

OPTION 3 ____________________

FAMILY PHONE TREE

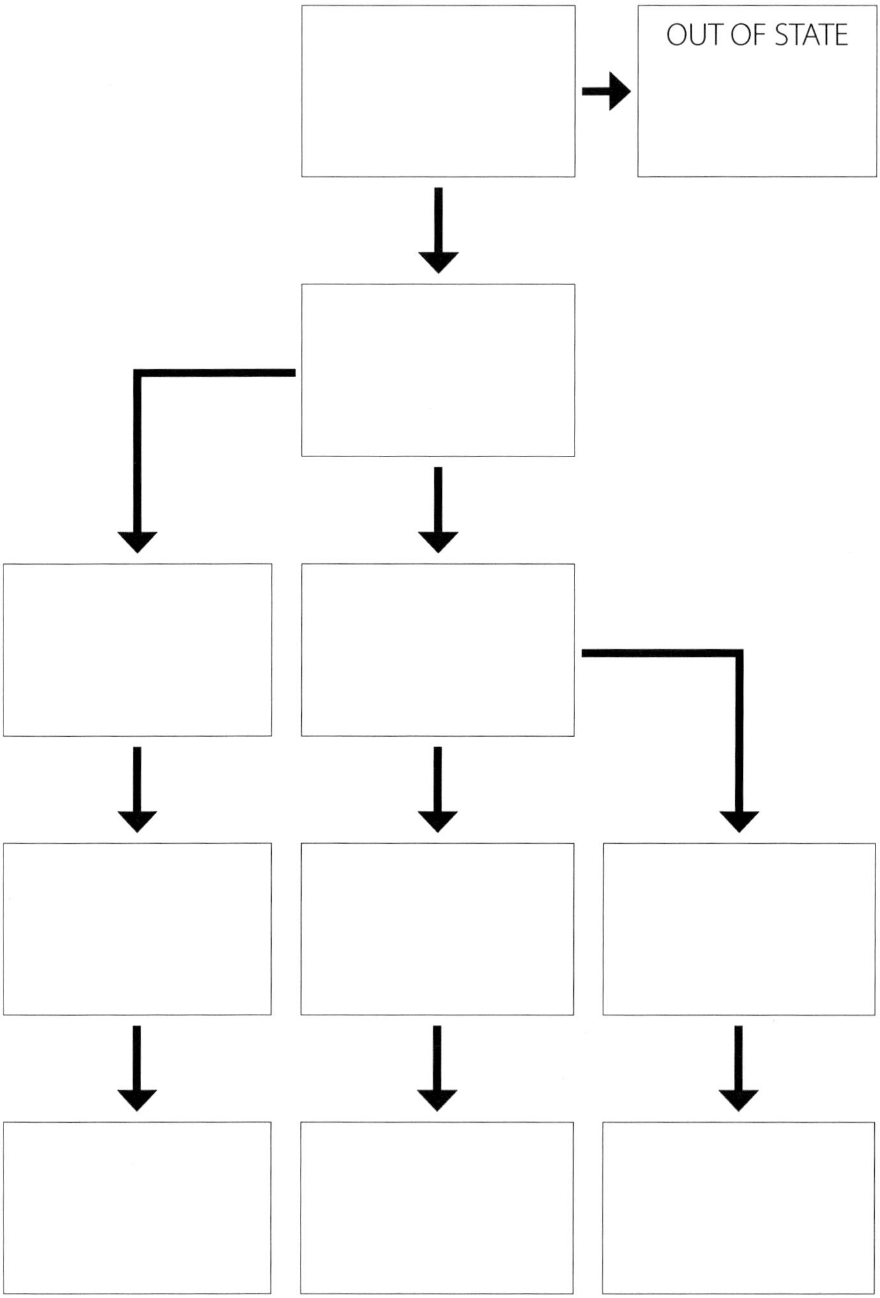

DIAGRAM OF OUR HOUSE

Show:
- Where the rooms are.
- Where each family member sleeps.
- Where the exit doors and windows are.
- Two(2) exits out of each room.
- Where pets are kept.
- Pre-planned meeting place outside.

⊗ Meeting Place

Draw a map of your home

SAFETY ZONE

Show distance from house to:

- Hospital
- Police Station
- Doctor's Office
- Designated Shelter
- Chosen Relative
- Selected Neighbor

● Our House

Draw evacuation route

DANGER PERIMETER

Show distance from house to:

- Tsunami Evacuation
- Flood Zone
- Rock or Mud Slide
- Nuclear Plant
- Hazardous Materials
- High Traffic Area
- High Crime Area

- Our House

PLAN OF GRANDPARENT'S HOUSE

Show:
- Where the rooms are.
- Where each family member sleeps.
- Where the exit doors and windows are.
- Two(2) exits out of each room.
- Where pets are kept.
- Pre-planned meeting place outside.

⊗ Meeting Place

Draw a map of your grandparent's house

SCHOOL EVACUATION PLAN

Show:
- Where the classroom is located.
- School's evacuation plan.
- Pre-planned meeting place outside.
- Alternate meeting place.

Draw a map of school

WORK EVACUATION PLAN

Show:
- Where you are located in workplace.
- Evacuation plan for building.
- Pre-planned meeting place outside.
- Alternate meeting place.

Draw a map of workplace

FAMILY SECRET PASSWORD

Note: Each child should have their own password. Teach your young children not to go anywhere unless they have your permission or, in an emergency, you give the person who picks them up the right password.

PASSWORD

CODE I-C-E

Note: Program numbers in your cell phone with the code ICE

If you are involved in an accident and have your mobile phone with you, Emergency Services will contact the numbers with the letters **ICE** next to the name.

If you're under 18, your **ICE** referral person should be a parent or an immediate member of your family authorized to make decisions for you. Other relatives or friends won't be able to make decisions for you if you're admitted to the hospital.

OUR CHILDREN'S FRIENDS

1

Name ______________________________

Address ______________________________

Home Phone ______________ Cell Phone ______________

School ______________________________

2

Name ______________________________

Address ______________________________

Home Phone ______________ Cell Phone ______________

School ______________________________

3

Name ______________________________

Address ______________________________

Home Phone ______________ Cell Phone ______________

School ______________________________

4

Name ______________________________

Address ______________________________

Home Phone ______________ Cell Phone ______________

School ______________________________

5

Name ______________________________

Address ______________________________

Home Phone ______________ Cell Phone ______________

School ______________________________

6

Name ______________________________

Address ______________________________

Home Phone ______________ Cell Phone ______________

School ______________________________

CHILD ID PROFILE

PHOTOGRAPH

Current Photo
(Head + Shoulder)

Date of Picture ____________________ Child's Age in Picture ____________

Personal & Medical Information

Keep in a safe and visible location

Full Name ____

Nickname ____

Street Address ____

City ____________________ State ______________ ZIP Code ____________

Social Security Number ____________________________________

Birth Date ______________________ Gender ______________________

Medication ____

Blood Type ____

Chronic Allergies/Illnesses ______________________________

Physician's Name ______________________________________

Physician's Phone Number ________________________________

PHYSICAL INFORMATION

Glasses____________________ Braces____________________

Piercings____________________ Race____________________

Hair Color____________________ Eye Color____________________

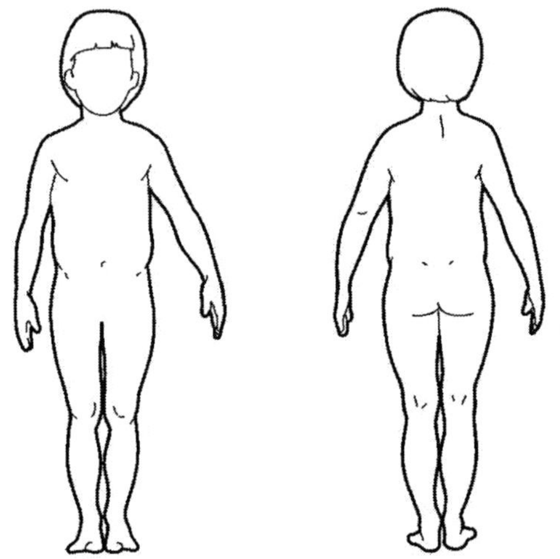

Indicate any identifying marks such as birth marks, moles, scars or previously broken bones, prosthetics or disabilities.

Instructions for Fingerprinting

Fingerprints are critical to a complete child identification record. The National Center for Missing & Exploited Children recommends that children be fingerprinted by competent, trained individuals, such as law enforcement personnel using appropriate ink and fingerprinting cards.

Right Thumb	Right Index	Right Middle	Right Ring	Right Pinky
Left Thumb	Left Index	Left Middle	Left Ring	Left Pinky

DENTAL CHART

Please have appropriate chart completed by your child's dentist at your next visit.

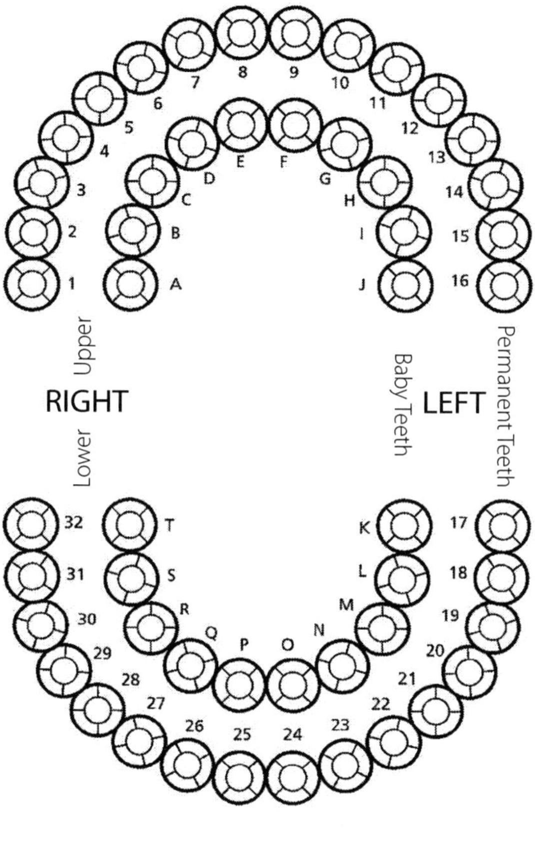

DNA Sample Instructions

1. **Hair Sample:** Remove 10 strands of hair from a brush that only your child uses. Make sure the roots and follicles are intact. Enclose them in the ***first*** zip-lock bag and write your child's name and date of sample on the white square. ***Store the bag in your freezer.***

2. **Cheek Swab Sample:** Make sure your child has not eaten within 1/2 hour. Take a Q-tip and swab the inside of the cheek for about 10 seconds. when finished, put the swab inside the ***second*** zip-lock bag and label with the child's name and date of sample. ***Store in your freezer.***

LOCATION OF EMERGENCY SUPPLIES

ITEM	LOCATION
Prescription medicine	
Water Supply	
Food Supply	
First Aid Kit	
Survival Kit	
Evacuation Kit	
Portable Radio	
Flashlight	
Batteries	
Fire Extinguisher	
Safety Ladder	
Smoke Detector	
Portable Generator	
Oxygen Tanks	
Insurance Records	
Financial Records	

OUR FAMILY **FIRST AID KIT**

1. ______
2. ______
3. ______
4. ______
5. ______
6. ______
7. ______
8. ______
9. ______
10. ______
11. ______
12. ______
13. ______
14. ______
15. ______
16. ______
17. ______
18. ______
19. ______
20. ______

OUR FAMILY **EVACUATION KIT**

1.

2.

3.

4.

5.

6.

7.

8.

9.

10.

11.

12.

13.

14.

15.

16.

17.

18.

19.

20.

OUR FAMILY **SURVIVAL KIT**

1.
2.
3.
4.
5.
6.
7.
8.
9.
10.
11.
12.
13.
14.
15.
16.
17.
18.
19.
20.

FAMILY AUTOMOBILES

1

Make ______________________ Type ______________________

Model Year ______________________ Color ______________________

Licence ____

2

Make ______________________ Type ______________________

Model Year ______________________ Color ______________________

Licence ____

3

Make ______________________ Type ______________________

Model Year ______________________ Color ______________________

Licence ____

OTHER TRANSPORTATION

Description:

FAMILY TRAVEL ITINERARY

AIRLINE INFORMATION

	FLIGHT #1	FLIGHT #2	FLIGHT #3	FLIGHT #4
Date				
Airline				
Airline contact number				
Flight Number				
Departure city				
Departure time				
Destination city				
Arrival time				

CAR RENTAL INFORMATION

Company	
Company contact number	
Confirmation number	
Pick-up date	
Pick-up location	
Drop-off location	
Drop-off date	

HOTEL INFORMATION

DATE	HOTEL	CITY/COUNTRY	RESERVATION #	CONTACT NUMBER

EMERGENCY IN FOREIGN COUNTRY

NOTE: Emergency Numbers may not be 911 in the country you are visiting. It is advisable to record the numbers in your book along with emergency words and phrases before you travel.

EMERGENCY NUMBERS

COUNTRY	AMBULANCE	POLICE	FIRE
1.			
2.			
3.			
4.			
5.			
6.			

EMERGENCY WORDS & PHRASES

ENGLISH	TRANSLATION
Help!	
Police	
Doctor	
Ambulance	
Hospital	
Fire	
Pharmacy	
Restroom	

BE PREPARED

- OFFICIAL CIVIL DEFENSE RADIO STATION:

____

- STATE CIVIL DEFENSE:

____

- COUNTY CIVIL DEFENSE:

____

- AMERICAN RED CROSS:

____

IMPORTANT CALL CENTERS

911 Emergency Calls: Police, Fire, Ambulance

711 Relay for Hearing Impaired

511 Traveler's Advisory

311 Police for Non-Emergency Matters

211 Community Services & Volunteer Opportunities

This book belongs to:

Family ____________________

Address ____________________

City ____________ State ____________ Zip ____________

Phone ____________________

IN AN EMERGENCY

Use The Necessary Information About Your Family Profile Recorded In This Book.

For 911— See Form F1 On Page 340.